Modern Ship Engineering, Design and Operations

Edited by Carlos Reusser

Published in London, United Kingdom

IntechOpen

Supporting open minds since 2005

Modern Ship Engineering, Design and Operations
http://dx.doi.org/10.5772/intechopen.94975
Edited by Carlos Reusser

Assistant to the Editor: Joel Perez Osses

Contributors
Prasadja Ricardianto, Imam Sonny, Céline Vaneeckhaute, Eleyadath Lakshmi, Velayudhan Sivanandan Achari, Machinchery Priya, Santiago Zárate González, Max F. Platzer, Garth V. Hobson, Sean P. Caraher, Paulo Santos Vasconcelos, Saulo Vasconcelos

First published in London, United Kingdom, 2021 by IntechOpen
IntechOpen is the global imprint of INTECHOPEN LIMITED, registered in England and Wales, registration number: 11086078, 5 Princes Gate Court, London, SW7 2QJ, United Kingdom
Printed in Croatia

British Library Cataloguing-in-Publication Data
A catalogue record for this book is available from the British Library

Additional hard and PDF copies can be obtained from orders@intechopen.com

Modern Ship Engineering, Design and Operations
Edited by Carlos Reusser
p. cm.
Print ISBN 978-1-83969-472-1
Online ISBN 978-1-83969-473-8
eBook (PDF) ISBN 978-1-83969-474-5

For EU product safety concerns:
IN TECH d.o.o., Prolaz Marije Krucifikse Kozulić 3, 51000 Rijeka, Croatia, info@intechopen.com or visit our website at intechopen.com.

Meet the editor

Carlos A. Reusser earned a BEng in Naval Electrical Engineering from the Naval Polytechnic Academy (APOLINAV), Chile, in 2001 and served in the Chilean Navy for twenty years. He also received a master's degree in Asset Management and an MSc and Ph.D. in Power Electronics, all from Universidad Tecnica Federico Santa Maria (UTFSM), Chile, in 2010, 2014, and 2020, respectively. From 2011 to 2019, Dr. Reusser was a lecturer at APOLINAV. In 2015, he became a lecturer at UTFSM. In 2020, he became an associate professor at Pontificia Universidad Catolica de Valparaiso, Chile. His main research interests include sensorless control, power converters, and electric ship applications.

Contents

Preface

This book integrates the technical knowledge obtained from educational and research activities to increase understanding of ship design, operation, and engineering. Modern ships are complex systems that involve structural optimization, energy generation and distribution, propulsion system design, ballast and wastewater management, and human resource administration.

The first section focuses on the design and operation of modern ships. It presents new approaches to ship propulsion systems based on high-performance sails whose aerodynamic characteristics approach those of rigid wings yet permit a reduction in the sail area in high wind and sea conditions. The design process of these sails is presented using computational analysis showing aerodynamic characteristics and optimum sail trim configurations. This section also discusses ballast wastewater, with a focus on ballast water treatment systems (BWTS) and ballast-free ships. New methods and standards in ship design are being developed to bring ecological balance and sustenance to marine ecosystems.

The second section discusses the logistics and management of modern ships. It presents several interesting and new approaches to maintaining ballasts and managing wastewater. In this context, many complex logistical issues must be overcome, whether in cabotage, maritime support, fluvial ships, or oil and gas rigs. The developed tools include organizing and forecasting tools designed to solve these problems in an organized, quick, and easy way.

Carlos Reusser
School of Electrical Engineering,
Pontificia Universidad Catolica de Valparaiso,
Valparaiso, Chile

Section 1

Modern Ship Design

Chapter 1

Aerodynamic Analysis and Design of High-Performance Sails

Sean P. Caraher, Garth V. Hobson and Max F. Platzer

Abstract

High-performance sails, such as the ones used on the America Cup boats, require sails whose aerodynamic characteristics approach those of rigid wings, yet permit a reduction in sail area in high wind and sea conditions. To this end, two-cloth sails are coming into use. These sails are constructed out of an articulated forebody that is a truncated ellipse, the aft of which has sail tracks, or rollers, along the edges to accommodate the twin sails. As the sails on either side need to be of the same length, due to the requirement to sail on different tacks, the two cloth sections need to be of equal length. The requirement then is to have their clews separated and able to slide over each other. More importantly, the transition between the rigid mast section and sails needs to be as aerodynamically smooth as possible in order to reduce drag and hence maximize the lift to drag ratio of the airfoil section that is made up of the mast and twin sails. A computational analysis using ANSYS CFX is presented in this chapter which shows that the aerodynamic characteristics of this type of two-cloth sail are almost as good as those of two-element rigid wing sections. Optimum sail trim configurations are analyzed in order to maximize the thrust production. Applications may soon extend beyond competitive sailing purposes for use on sailing ships equipped with hydrokinetic turbines to produce hydrogen via electrolysis (energy ships). Additionally, high performance sails can be used onboard cargo ships to reduce overall fuel consumption.

Keywords: energy ship, sail aerodynamics, twin-skin sail, computational fluid dynamics, high performance sails

1. Introduction

In his book "The 40-Knot Sailboat" [1] Bernard Smith, the former director of the United States Naval Weapons Laboratory in Dahlgren, Virginia, gives a very illustrative history of the sailing ship development over the past five millennia. It is quite apparent that throughout this long history the cloth sail was considered the obvious best choice because of its ease to adjust to fast changing wind conditions. It is also apparent that the sail aerodynamics remained poorly understood before the pioneering insights of Kutta and Joukowski about the principles of lift generation in the early 1900s. At about this time interest in the sailing ship as a commercially or militarily important technology waned due to the transition to the steamship. Since then, the sailboat is largely regarded as a vehicle of interest only to competitive and recreational sailors. It is therefore not surprising that the cloth sail remained to be regarded as the logical tool to be used for lift, or thrust, generation.

In the early days of aeronautics the very thin foil was also regarded as the most plausible lift generator, inspired by the observation of bird flight. Lilienthal and the Wright brothers pioneered its use and it remained in use until 1917 when the Dutch airplane builder Anthony Fokker showed that relatively thick airfoils provided the German fighter airplanes with better maneuverability in WWI. Since then, their use has become standard practice in aeronautical engineering.

When Bernard Smith began his search for greatly increased sailing speeds in the 1950s he started with the observation: "After centuries of struggle, the fastest sailboats of our time, whether clipper ships, America's Cup racers, inland lake scows or the amazing double-hulled canoes of the Pacific Islanders, are, after all, only a little faster than the speediest vessels Magellan saw in his day". He recognized the need for superior aerodynamic sail characteristics offered by modern airfoils and similar characteristics offered by modern hydrofoils. His pioneering developments ultimately led to the "sail rocket" which achieved speeds exceeding 60 knots.

High-performance sails, such as the ones used on the America Cup boats, require sails whose aerodynamic characteristics approach those of rigid wings, yet permit a reduction in sail area in high wind and sea conditions. To this end, two-cloth sails are coming into use. These sails are constructed out of an articulated forebody that is a truncated ellipse, the aft of which has sail tracks, or rollers, along the edges to accommodate the twin sails. As the sails on either side need to be of the same length, due to the requirement to sail on different tacks, the two cloth sections need to be of equal length. The requirement then is to have their clews separated and able to slide over each other. More importantly, the transition between the rigid mast section and sails needs to be as aerodynamically smooth as possible in order to reduce drag and hence maximize the lift to drag ratio of the airfoil section that is made up of the mast and twin sails. Applications may soon extend beyond competitive sailing purposes for use on sailing ships equipped with hydrokinetic turbines to produce hydrogen via electrolysis (energy ships), as proposed in Ref. [2]. In this application it will again be very important to maximize the sail thrust while minimizing the ship drag by means of hydrofoils.

A computational analysis using ANSYS CFX is presented in this chapter which shows that the aerodynamic characteristics of this type of two-skin mainsail are almost as performant as those of two-element rigid wing sections [3]. Optimum sail trim configurations are analyzed in order to maximize the thrust production.

2. Modern sail aerodynamics

The use of sails on large vessels, including cargo ships and tankers, is not a new idea. It has been proposed countless times and many concepts for sail assisted vessels have been proposed [3]. The use of sails may serve to lower carbon emissions from large scale shipping or even be used to harvest energy using hydrokinetic turbines, see References [2, 4]. Generally, the proposed concepts make use of either traditional cloth sails or articulating wing sails. However, the latest edition of the America's Cup, the most technological advanced sailing competition, may have pioneered another sail configuration that could provide the usability advantages of cloth sails and the performance gains of a rigid wing sail.

The America's Cup has long been the pinnacle of high-performance sailboat design. Ever since the cup was first competed for in 1851, by the 'radical' looking schooner *America*, the race has produced innovations in high performance sailboat design. Rigid wing sails were first introduced to the Cup in 1988 by Dennis Connor's syndicate. More recently wing sails have been used onboard the AC72 and AC50 catamarans [5]. These rigid wing sails are composed of multiple elements that can

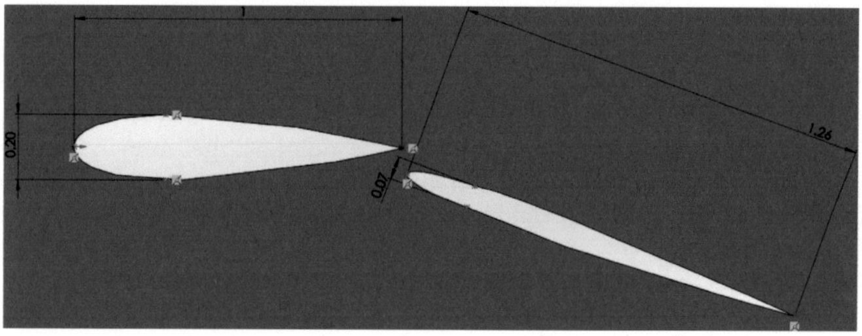

Figure 1.
Possible configuration of a rigid wing sail [6].

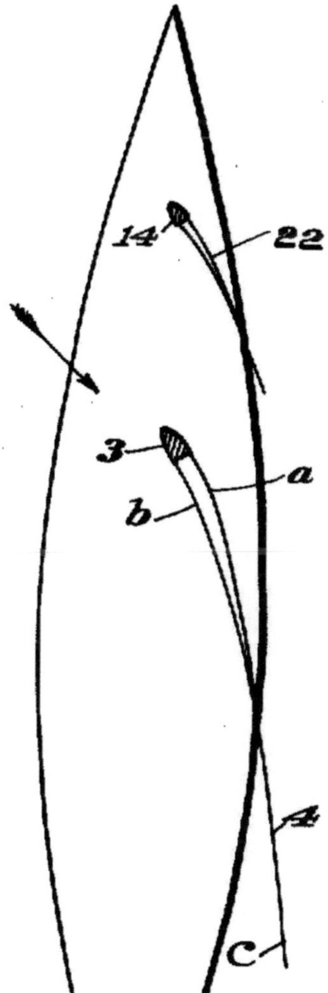

Figure 2.
Drawing from original Herreshoff patent [8].

be articulated to create efficient multi element airfoil sections. **Figure 1** shows the multi-element rigid wing section tested by Johnson. These sails seemed to produce incredible performance for both the AC72 and AC50 classes, however they are held back by the difficulty of handling a rigid wing. For safe storage these large wings must be taken down in order to prevent them from being damaged by winds while not being used. Their construction generally is very light using an inner carbon structure with a mylar covering. The process of taking one of these sails down took 30 to 40 people approximately one hour with the aid of a large crane in the case of the AC72 wing [7]. In addition, unlike traditional cloth sails, rigid wing sails cannot be reduced or reefed as wind speed increases, which could leave a vessel in a dangerous situation in adverse weather conditions.

These difficulties as well as the increased cost and technical complexity pushed America's Cup organizers to specify the use of a double skin mainsail for the 36th America's Cup. Twin-skin mainsails are not a new concept, however. The idea was first filed for patent by the famous sailboat designer Lewis Herreshoff in 1925, shown in **Figure 2** [8]. This design uses two cloth mainsails that are attached to an elliptic mast section to create an airfoil like shape with finite thickness. The aerodynamic performance of this sail configuration is mostly unknown because most of the development of twin-skin mainsails was done in secrecy by teams competing in the 36th America's Cup. However, these designs seem to promise greater performance than traditional cloth sails without the hassle of a fully rigid wing.

In order to get estimates of the performance that can be expected from these twin skin mainsail sections, a CFD study was conducted on a representative two-dimensional twin skin mainsail section. The analysis was conducted two dimensionally using ANSYS CFX software. The section selected was designed to represent what a twin skin mainsail may look like when hoisted. This chapter will present the results of this modeling and the challenges experienced while attempting to accurately predict the aerodynamic characteristics of the twin-skin mainsail.

3. Discussion of the geometry

The section was designed around the use of a two-to-one elliptic mast section that would serve as the leading edge of the mast. In practice this mast section would be designed to rotate in order to be able to present a smoother airfoil like section for varying angles of incidence. This technique is already used onboard high-performance sailboats and has proven its feasibility in numerous circumnavigations. The sails are then connected to the mast section on either outboard edge. Each sail is of identical chord length so that the configuration can be articulated to accommodate sailing on either tack. In order to induce camber in this section while imposing the condition that each sail is of equal length, the trailing edges are designed to slide over one another. This artificially allows the leeward side of the setup to be moved towards the mast inducing camber into the suction side of the sail. This design is shown

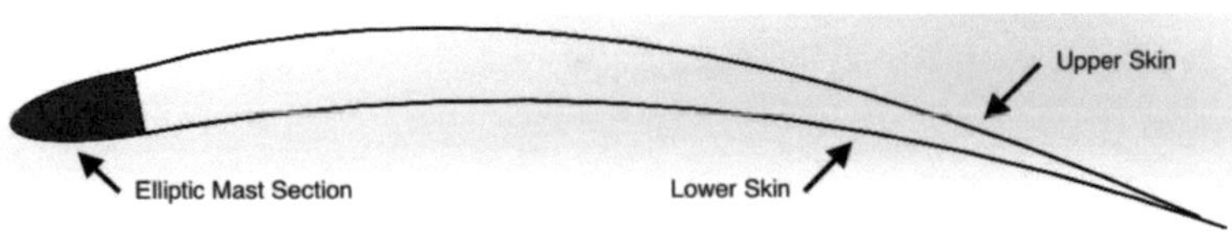

Figure 3.
Twin-skin mainsail geometry.

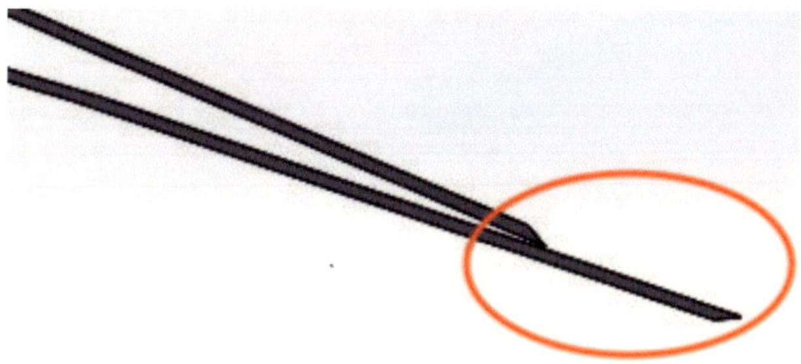

Figure 4.
Trailing edge diagonal cut treatment for simplified meshing.

in **Figure 3**. For ease of meshing in CFD, the trailing edge treatment shown in **Figure 4** was used. This close up also shows in detail how the upper sail is allowed to move forward to induce camber and account for mast rotation.

4. Domain enlargement study

The nature of the boundary conditions in CFD simulations requires the edges of the computational domain to be sufficiently far away from the object being tested. To find a domain size, a simple square domain was created around the proposed geometry. Domain size was slowly increased, and aerodynamic coefficients were monitored. To minimize computational time a small range of angles of attack were chosen for analysis at each domain size. The model was run with inflation layers clustered around the sail providing a non-dimensional distance of the first grid from the sail (Y^+) of approximately one across the entire sail. Y^+ is a non-dimensional distance from wall boundary conditions calculated based on turbulent skin-friction on the wall. For accurate resolution of boundary layer affects, Y^+ should be in the single-digits. For this study the turbulent kinetic energy and dissipation (k-e) turbulence modeling was used to provide a fully turbulent solution.

As domain size was increased (**Figure 5**), measured as the distance to the boundary from the sail, the lift coefficient begins to asymptote as the boundary distance reaches 70 meters, shown in **Figure 6**. In addition, the variation of the vertical velocity along the top of the domain (**Figure 5**) was deemed to be sufficiently small and is shown in **Figure 7**. This leads to a 140 m by 140 m, or

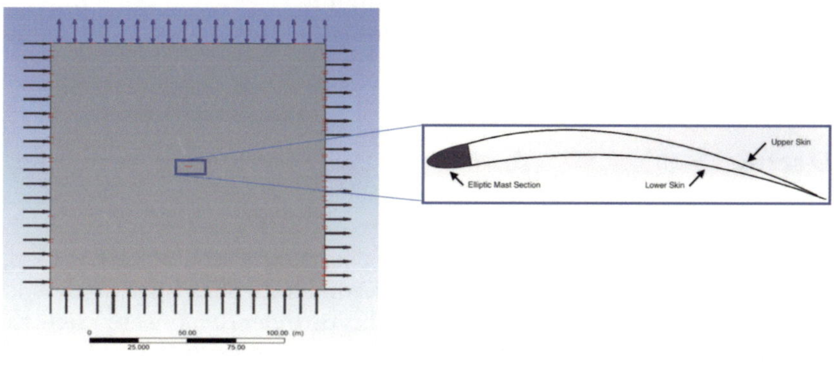

Figure 5.
Configuration of boundary conditions.

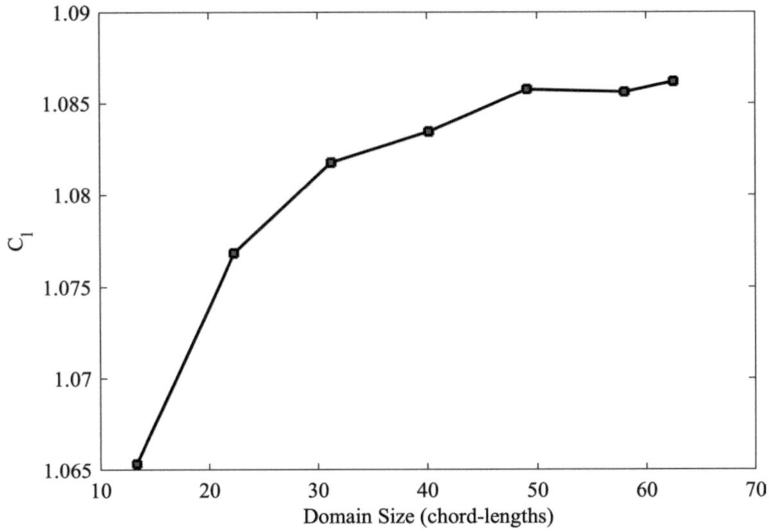

Figure 6.
Change in lift coefficient at 2 degree angle-of attack, Re = 2,000,000, as domain size is increased.

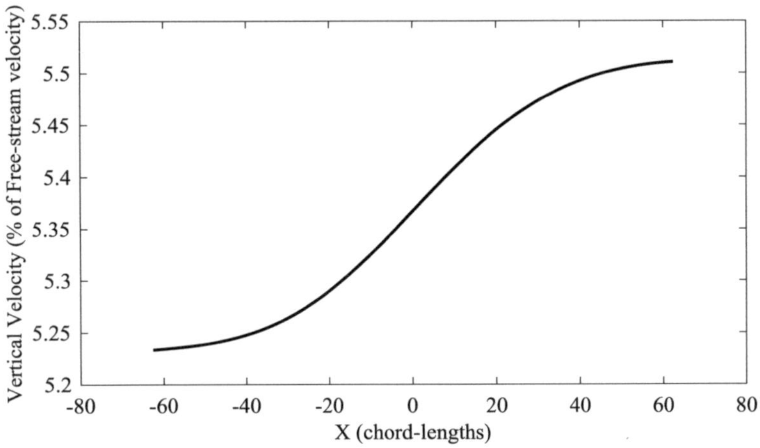

Figure 7.
Variation of vertical velocity on the entrainment boundary for 2 degree angle-of-attack.

approximately 60 chord-lengths, domain around the twin-skin sail. From this exploration, it was decided that all future simulations would be conducted with this domain size.

5. CFD setup

From the domain size determined, two different CFX jobs were created. The only difference between them was turbulence modeling. The first job used k-epsilon turbulence with scalable wall functions. The second used Shear Stress Transport model (SST) with Gamma Theta transition. This turbulence model predicts the transition of flow from laminar to turbulent using turbulent kinetic energy and vorticity. The boundary conditions are also shown in **Figure 5** (annotate, Inlet, Bottom, Top, Outlet). The angle of attack was specified by changing the u and v velocity at the inlet boundary condition, which was spread across two faces. An entrainment boundary condition was located at either the top or the bottom of the domain depending on the angle of attack to allow for circulation affects. Further details regarding this research can be found in Caraher's thesis available through the Naval Postgraduate School, Monterey, California [9].

6. Results

The results from the two different models show stark differences in how aerodynamic coefficients were estimated and how the flow fields behaved. The Shear Stress Transport (SST) turbulence model and Intermittency Momentum Thickness (g-q) transition model struggled to resolve the flow field near stall, shown in **Figure 8** by the inconsistent calculation of lift coefficient above 12 degrees angle of attack. The SST model is a blending of the k-w (vorticity) and k-e turbulence models, with the k-w equations being solved near the wall. In comparison the k-e fully-turbulent solution shows a benign stall. This stall begins at the trailing edge as slowly works forward as angle-of-attack is increased, shown in **Figure 9** for the k-e case.

The two models produce different predictions of drag especially at lower angles of attacks, see **Figure 10**. The transition model shows a laminar drag bucket between −5 and 5 degrees angle-of attack. This drag bucket was unexpected and the sharp increase in drag shown by the transition model occurs over a quarter degree change in angle-of-attack.

This drag bucket is caused by the transition models resolution of laminar flow on the upper surface on the twin-skin mainsail at these angles of attack. This is shown by examining the pressure and skin friction coefficients. Both models predict similar pressure coefficients both in the drag bucket and at higher angles of attack, shown in **Figures 11** and **12**. However, skin-friction coefficients differ within the region of the drag bucket, plotted in **Figure 13**. This plot shows that laminar flow on the lower surface transitions relatively early on the twin-skin mainsail, shown by a sharp increase in skin-friction coefficient as the flow transitions. This area of transition is visible in the flow field as a separation bubble, shown in **Figure 14**. From this point aft, the skin friction coefficients predicted on the lower surface by each model are

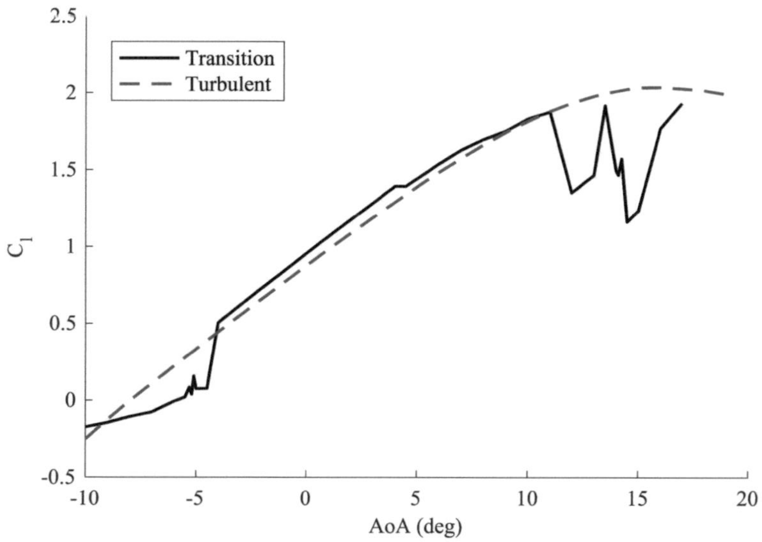

Figure 8.
C_l comparison between SST and fully turbulent at Re = 2,000,000.

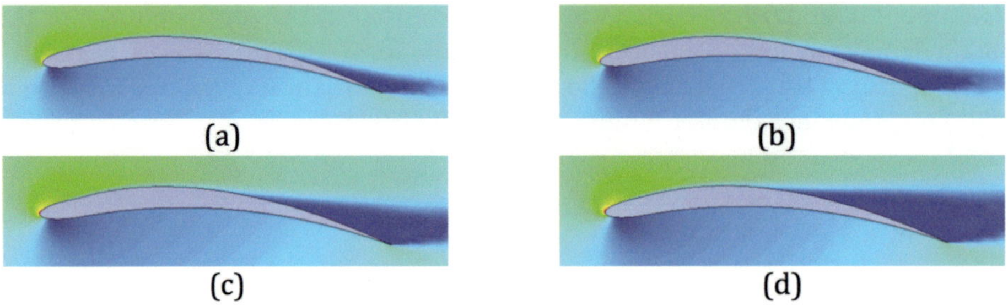

Figure 9.
Growth of trailing edge separation in K-epsilon model stall. (a) 13 degrees, (b) 15 degrees, (c) 17 degrees, and (d) 19 degrees.

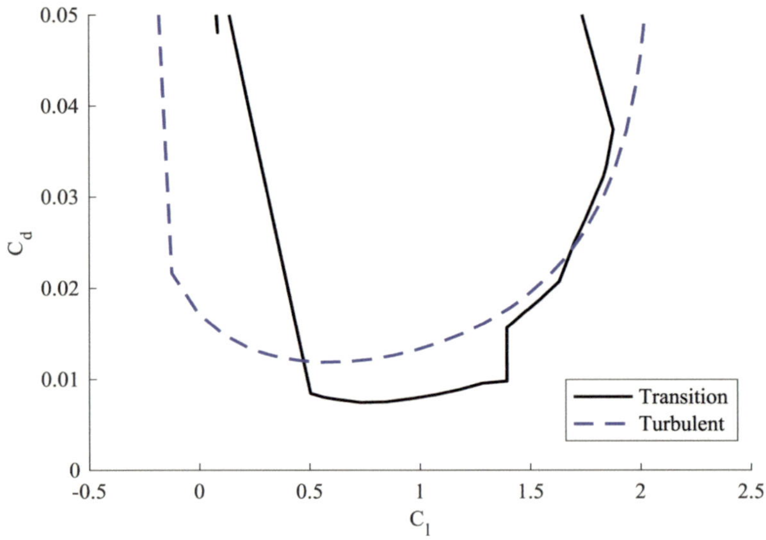

Figure 10.
C_d comparison between SST and fully turbulent at Re = 2,000,000.

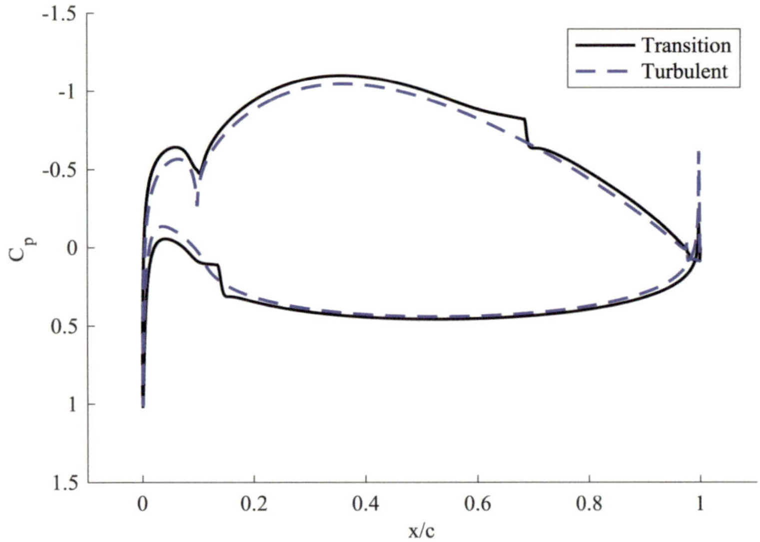

Figure 11.
Surface pressure coefficients at AoA = 1 degree, Re = 2,000,000.

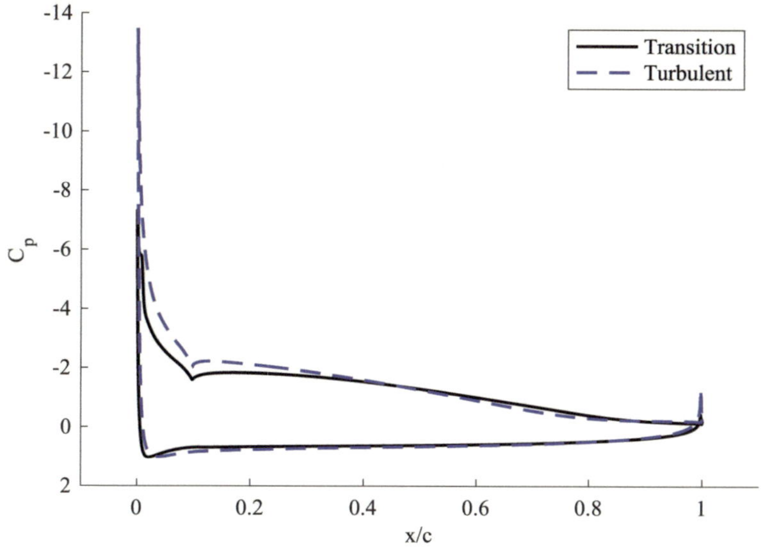

Figure 12.
Surface pressure coefficients at AoA = 10 degrees, Re = 2,000,000.

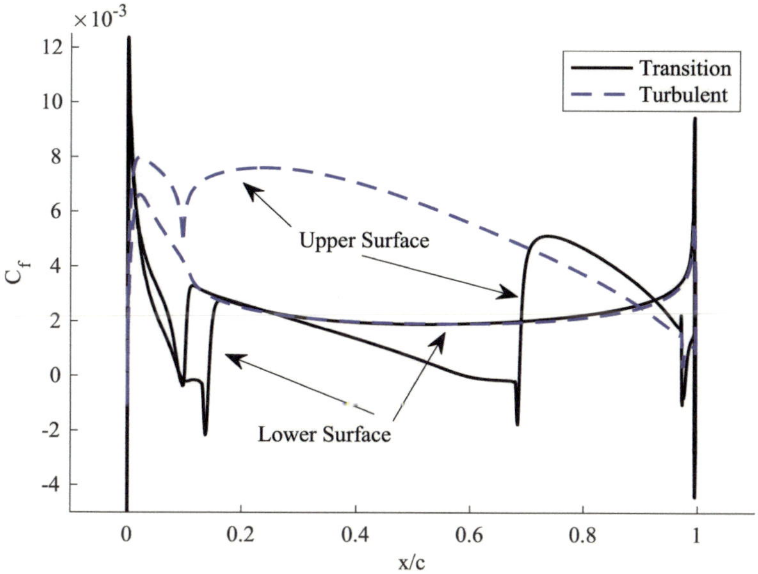

Figure 13.
Surface friction coefficients at AoA = 1 degree, Re = 2,000,000.

nearly identical. However, the SST model does not predict transition on the upper surface until 70% chord within the region of the drag bucket. This results in much lower skin-friction coefficients on most of the upper surface, shown in **Figure 13**.

At higher angles of attack, laminar flow is no longer predicted along much of the upper surface. In this region, SST predicts transition as the flow accelerates around the leading edge of the mast section, shown as a separation bubble in **Figure 15**. SST's prediction of early transition means that most of the flow around the twin-skin mainsail at higher angles of attack is turbulent, and therefore both models predict similar skin friction coefficients, shown in **Figure 16**. This results in closer prediction of drag coefficients between the two models above 5 degrees angle of attack.

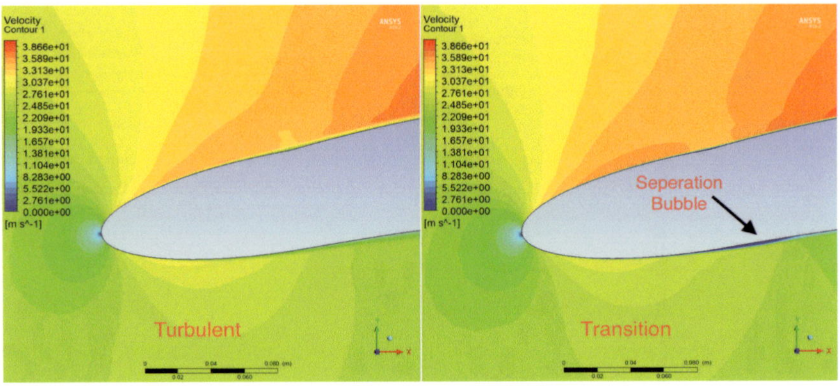

Figure 14.
Leading edge velocity field at 1 degree, Re = 2,000,000.

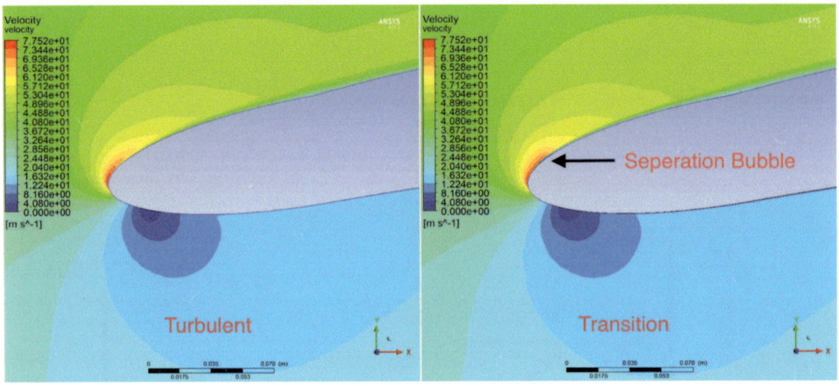

Figure 15.
Leading edge velocity field at 10 degrees, Re = 2,000,000.

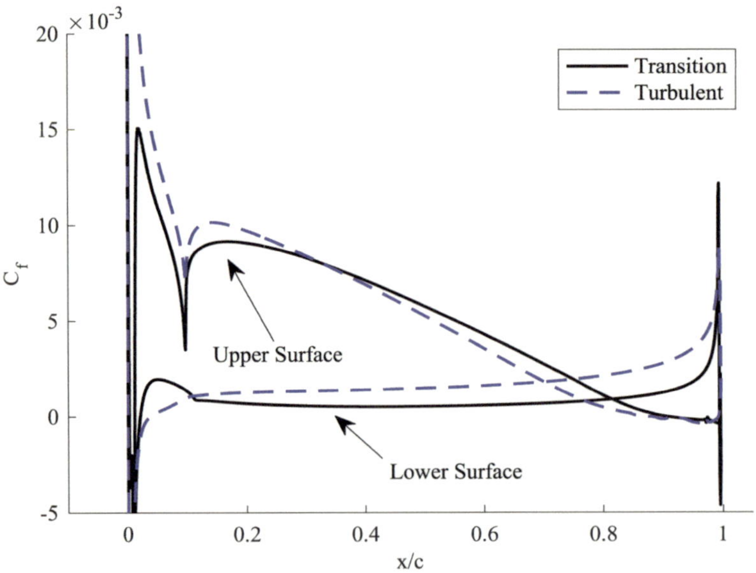

Figure 16.
Surface friction coefficients at AoA = 10 degrees, Re = 2,000,000.

The environment in which the twin-skin mainsail will be operating dictates that the fully turbulent solution will likely be closer to actual performance. This is due to the constantly changing flow-field that sails experience when deployed. Variations in wind speed and movement of the vessel mean that the velocity and angle of attack that a sail experiences is also constantly varying. This variation will suppress the formation of laminar flow as the chaotic flow-field should favor a turbulent boundary layer. In addition, these models were run without the influence of surface roughness. A real sail will have finite roughness caused by cloth texture as well as imperfections in the surface caused by seams. This should serve as a second factor that should promote turbulent flow around the twin-skin mainsail when deployed.

7. Conclusions

The ANSYS CFX analysis of two-dimensional flow past twin-skin mainsails presented in this paper yielded the following major results:

a. The sail is able to produce lift coefficients up to a maximum of 2.0

b. The drag coefficient predictions vary significantly depending on the choice of turbulence and transition modeling. This was to be expected. Nevertheless, a low drag region is predicted in either case between lift coefficients of zero to 1.4.

c. The twin-skin sail presents the ANSYS CFX analysis with a greater than usual challenge because of the slope discontinuities caused by the transition from the elliptic leading edge to the upper and lower skins and on the upper surface near the trailing edge.

d. In a previous analysis of the NACA 0012 airfoil [6] the code produced a remarkable agreement with the experiment in the low angle of attack range, thus giving confidence in its ability to predict transitional flows.

e. The prediction of separation bubbles and the onset of stall requires further detailed study. Fully turbulent calculations predict a rather benign trailing edge stall. If validated in future computational and experimental investigations this feature will be very welcome.

The CFD data that is presented has not been validated by comparison to known data sets. Publicly available data sets concerning the performance of twin-skin mainsails do not exist. These data sets may exist within internal team documentation for the 36th America's Cup, but due to the competitive nature of the event, teams have not published their findings. Despite this there is high confidence that the performance estimations presented in this paper are accurate due to validation of CFX code by Johnson [6].

It appears likely that the twin-skin mainsail will find further application in highly competitive sailing competitions, such as the America's Cup race. In addition, another application may occur in the operation of autonomous sailing ships equipped with hydrokinetic turbines and electrolyzers to produce hydrogen. As explained in Ref. [1], such energy ships require highly efficient sails to produce the propulsive power necessary to overcome the turbine drag and maximize energy production.

When deployed on an ocean-going vessel twin-skin sails should offer an advantage over rigid sails because of ease of stowage and operation. Cloth twin-skin sails can be stowed in either the mast or boom section by rolling the cloth within these sections, unlike rigid sails that cannot easily be stowed. This provides a distinct advantage when operating away from shore in extreme weather and sea-states. By optimizing the sail's performance and aerodynamics, it facilitates the overall system optimization including path planning. In the case of a sail assisted cargo vessel, path planning will consist of optimizing the vessels route to take advantage both weather and sea conditions to minimize fuel consumption.

A detailed investigation into the two-dimensional aerodynamics of a twin skin sail has been completed. The simulations included the effect of computational domain size upon the induced circulation around the airfoil. Additionally, both fully turbulent boundary layer flow as well as transitional flow was investigated. It was hypothesized that sailing ships will likely experience fully turbulent flow over most of the sail due to surface roughness and unsteady flow hence these simulations were most realistic.

Based on the success of the most recent America's Cup competition, twin-skin cloth sails appear to be the most suited to high performance as well as ease of use both in raising and lowering the sails. Hence this concept could be used on large ocean-going ships for either primary propulsion or as auxiliary propulsion to reduce overall fuel burn during transit.

Acknowledgements

The authors gratefully acknowledge helpful comments and advice from Professor Anthony Gannon, Naval Postgraduate School, Monterey.

Conflict of interest

The authors declare no conflict of interest.

Author details

Sean P. Caraher, Garth V. Hobson* and Max F. Platzer
Naval Postgraduate School, Monterey, USA

*Address all correspondence to: gvhobson@nps.edu

References

[1] B.S. Smith, "The 40-Knot Sailboat", Echo Point Books and Media, LLC, 1963

[2] M.F. Platzer and N. Sarigul-Klijn "The Green Energy Ship Concept: Renewable Energy from Wind over Water". Springer Briefs in Applied Sciences and Technology, Springer 2021

[3] C. Delbert, "This Wind-Powered Super Sailboat Will Carry 7,000 Cars Across the Atlantic," Popular Mechanics, Oct. 07, 2020. https://www.popularmechanics.com/science/energy/a34272175/wind-powered-sailboat-cargo-shipping-future/ (Accessed May 20, 2021).

[4] N. Gardner, "Brief guide to sail-assisted cargo ships," Thetius, Sep. 11, 2020. http://thetius.com/brief-guide-to-sail-assisted-cargo-ships/ (Accessed May 15, 2021).

[5] A. Cup, "History of the AMERICA'S CUP," 36th America's Cup presented by PRADA. https://www.americascup.com/en/history (Accessed May 06, 2021).

[6] R. Johnson, Susan, Computational aerodynamic analysis of airfoils for energy-producing sailing ships. Calhoun, 2020. [Online]. Available: https://calhoun.nps.edu/handle/10945/65559

[7] J. Griffin, "AC75 Double Luff Mainsail," Sail-World, Aug. 27, 2018. https://www.sail-world.com/news/209295/AC75-Double-Luff-Mainsail (Accessed May 06, 2021).

[8] L. F. Herreshoff, "Sailboat," US1613890A, Jan. 11, 1927.

[9] P. Caraher, Sean, Computational analysis of twin-skin cloth sails for high performance sailing vessels. Calhoun, 2021. [Online].

Chapter 2

Systems and Operation of Ballast Water in Ships with the Changing Ballast Water Management Policy

Eleyadath Lakshmi, Machinchery Priya and Velayudhan Sivanandan Achari

Abstract

Bio-invasion caused due to ballast water discharge is one of many problems in marine pollution. Countries such as Canada, Brazil, USA and Australia recognized the problems associated with ballasting and deballasting. Countries affected with invasive species formulated specific laws for discharging ballast water in their respective ports. Under the coordination of IMO, countries came together and stressed for globally accepted guidelines that each and every ship has to comply with, while entering any port. In the wake of this, IMO in a convention (2004) on ballast water, proposed guidelines for performing proper ballast water management. This includes ballast water exchange, ballast water treatment, port reception facility, technology approval process, sampling ballast water, analysis methods of ballast water and risk assessment in the convention. Eventually the 2004 convention was found to be inadequate in providing complete elimination of bio invasion. Amendments are made to the 2004 convention over the years for ballast water management. It is found that the member states should share technology among developing countries in establishing sampling and testing laboratories. Region specific sampling analysis and research has to be formulated to understand the bio-invasion based on region and characteristics of different target species in evaluating risk assessment. The D2 standard mentioned in the 2004 convention should be changed from size specific to 'no organism' standard in ballast water for discharge. New combination of BWT systems and 'no ballast' system with modification to the ship design should be tested, developed and implemented to bring in ecological balance and sustenance in the marine ecosystems.

Keywords: ballast water, ballast water treatment, management policy, guidelines, systems and operations

1. Introduction

The ballast literally meaning "any material that is used to balance an object to maintain its buoyancy". Ships need ballast water to maintain its stability and maneuverability when she is empty or partially loaded. The water in ballast is adjusted continuously by the crew based on the design and weather conditions of the sea through which the ship is navigating. The quality of water and the organisms present in the ballast is primarily determined by the route and the region through which the ship is

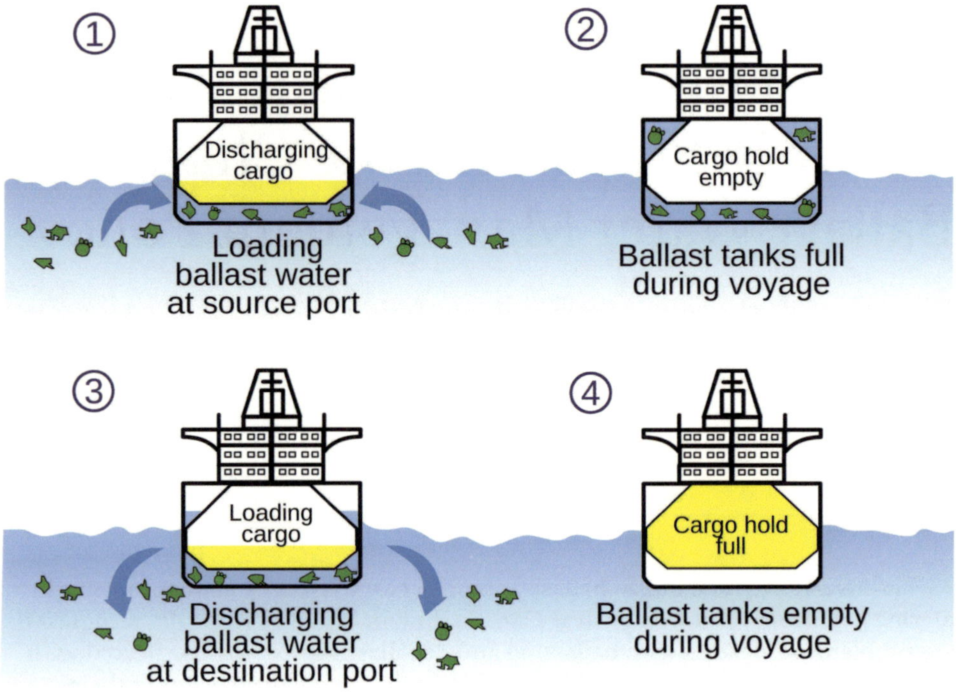

Figure 1.
Pictorial representation of ballast water. Source: MaxxL Derivative work: Thorsten Hartmann - [File:Water] File:Water pollution by ballast water de.svg, CC BY-SA 3.0, https://commons.wikimedia.org/w/index.php?curid=33556135.

traveling (**Figure 1**). Some organisms such as comb jelly fish, chinese mitten crabs, European green crabs, mussels, whelks, holoplankton, American jack knife clam and vibrio cholera has survived the harsh conditions inside the ballast tanks during long voyages. These organisms have established themselves in different environment when released causing disruption to the water quality and ecology of the respective ports. The ability of planktons, microbes, and pathogens to pump into ship's ballast system and survive relatively long voyages, drifting in the ballast water till the end of the voyage was identified as early as 1897 [1]. In 1904 scientist first recognized the signs of invasive species after a mass occurrence of the Asian phytoplankton algae *Odontella (B. sinensis)* in the North Sea [2]. With the growing awareness on protection and conservation of environment, United Nations held a conference on Human Environment in 1972 and declared the necessity for safeguarding the resources and environment. By 1973 IMO adopted an International convention for prevention of pollution from ships due to operational and accident causes (MARPOL-Marine Pollution). The protocol was later adopted in 1978 after numerous tanker leakages occurred during 1976–1977. Initially the ballast water and sediment management was first categorized under the MARPOL. Later on, IMO realizing the importance of ballast water management considered it specifically. under a separate category.

By 1980s, Canada and Australia were among the first countries in realizing the problems associated with invasive species and was brought to the attention of the international community. United Nations Conventions on the Law of the Sea, (UNCLOS) 1982 gave directions and stressed on the need for all the states to prevent, reduce, and control accidental or intentional introduction of species into the marine environment and to prevent, reduce, and control pollution of the marine environment from any source. By 1990, IMO created a separate working group within the marine environment protection committee (MEPC) to investigate the impact of ballast water operations in ports. Initially, there were ambiguity on whether to categorize ballast water discharge as marine pollution. Finally, it was construed to consider it under marine pollution in the UNCLOS. According to the convention, marine pollution means "the introduction by man, directly or indirectly of substances or energy into the marine environment, including estuaries, which results or is likely to result in such deleterious effects as harm to living resources and marine life, hazards to human health, hindrance to marine activities, including fishing and other legitimate uses of the sea, impairment of quality for use of sea water and reduction of amenities". UNCLOS provisions were found to be more effective in safe guarding the rights and responsibilities of the port rather than providing solutions to the bio-invasions and pollution caused by ballast water. The AGENDA 21 of the Rio declaration (1992) which call on nations to consider regulation of ballast water discharge to prevent the spread of non-indigenous organisms and advocates on the precautionary and polluter pays principle which could bring in more nations and ship authorities accountable in protecting the marine biodiversity, prevention of bio invasion and marine pollution by the discharge of ballast water.

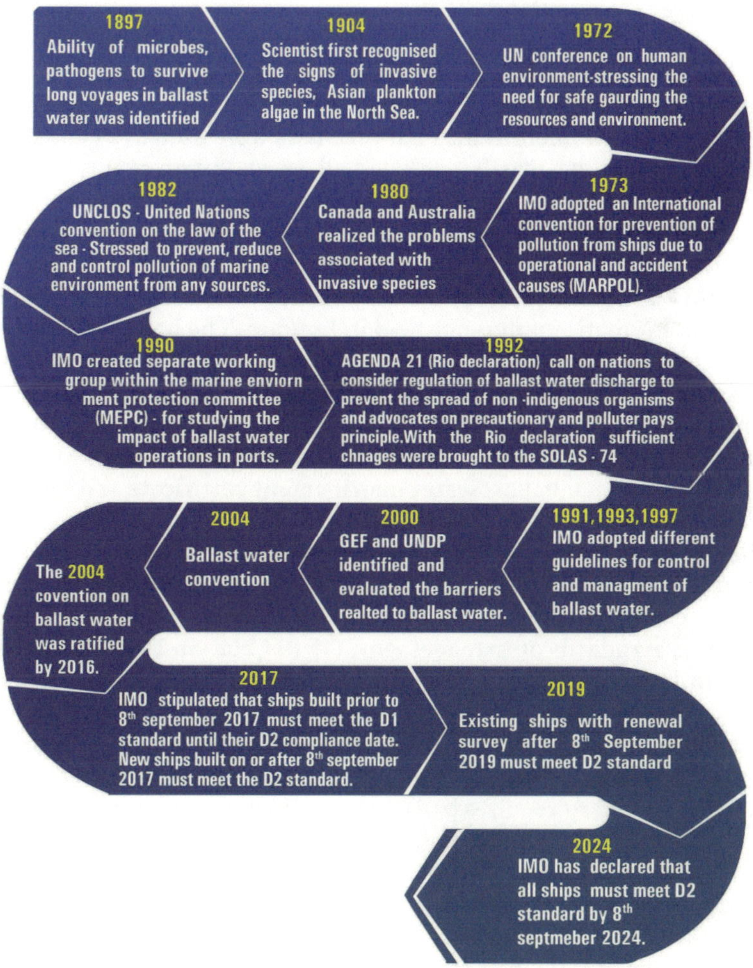

Figure 2.
Sequence of events in ballast water regulations.

SOLAS 74 – safety convention and ISM code which give guidelines for stability and safety operations, is applicable to the ballast water management in ships. With the convention on biological diversity at Rio Declaration, 1992, sufficient changes had been brought into the SOLAS 1974 and ISM code to be implemented for safe operations of ballast water. In 1991, 1993 and 1997, IMO adopted different guidelines for the control and management of ballast water to minimize the transfer of harmful and pathogenic aquatic organisms. But by 1997 with the zebra mussel invasion in US and Canada, the United Nations General assembly passed a resolution to prevent ballast water pollution. Guidelines were modified for better ballast management practices which the states could adopt by means of their national legislation [3, 4]. The guidelines gave directions to the nations in framing and enacting domestic laws for minimizing and reducing the risks associated with the ballast water discharge. A joint initiative by IMO (2000) named Globallast was launched in association with the Global Environment Facility (GEF) and UNDP to identify and overcome barriers related to ballast water effectively. The sequence of events in ballast water regulations is depicted in the **Figure 2**.

2. Ballast Water Convention 2004

General guidelines for ballast water was not giving way for a solution to the unique problems such as bio invasion and marine pollution created by ballast water intake and discharge. Numerous international laws were used as a general guideline for countries performing ballasting and de-ballasting. But no country could clearly give directions on how to manage ballast water properly. Since ballast water is a global concern, separate rules and regulations for each nation will not bring any consensus among different states for ballast water discharge.

IMO in 2004 by consensus adopted the Ballast Water Management Convention (BWMC) at a diplomatic conference held in London. The ballast water working group of IMO (2004) drafted ballast water discharge standards which aims at preventing and eliminating ballast water pollution. It comprises of 22 articles, regulations and 1 annexure detailing general obligations of states to implement the technical requirements. It has an appendix setting model formats for the issuance of international ballast water management certificate and ballast water record book.

To achieve the goal of elimination and prevention of bio-invasion the BWM Convention required all vessels to implement a Ballast Water Management Plan and adhere to clearly defined management standards. The BWM Convention established two standards of management: (1) Regulation D-1, (the ballast water exchange standard), and (2) Regulation D-2, (the ballast water performance standard).

Regulation D-1 required a minimum ballast water exchange volume of 95%, while regulation D-2 established a concentration threshold for ballast water discharge. The D-2 standard requires ballast water discharge to contain: 1) Less than 10 viable organisms per cubic meter greater than or equal to 50 mm in minimum dimension. 2) Less than 10 viable organisms per milliliter less than 50 mm in minimum dimension and greater than or equal to 10 mm in minimum dimension. 3) Less than the following concentrations of indicator microbes, as a human health standard: 1) Toxigenic *V. cholerae* (O1 and O139) with less than 1 colony forming unit (cfu) per 100 ml or less than 1.0 cfu per 1.0 g (wet weight) zooplankton samples. 2) *E. coli* less than 250 cfu per 100 ml. 3) Intestinal Enterococci less than 100 cfu per 100 ml.

2.1 Ballast Water Management and Ratification with the Convention

The BWM Convention came into force 12 months after ratification by 30 States representing not less than a combined 35% of the world's merchant shipping gross tonnage. Regulations D-1 and D-2 will be implemented on a phased schedule based on age and ballast water capacity of each vessel, with all vessels eventually required to meet the D-2 standard [5].

There is no universally applicable currently available method for shipboard treatment of ballast water. This creates the space for a diverse research and development in the ballast water management. A more consolidated effort has to be put in for developing an effective ballast water management. Initially, community composition of the ballast water in ships coming to ports around the world were documented and recorded. Gradually, with the growing awareness on marine environment, different studies were being focused on the tolerance levels of species transported by ships to different ports. There is currently insufficient data to confidently quantify the probability of invasion associated with any particular inoculum density (or discharge standard). As a result, laboratory, field and modeling studies examining the relationship between invasion risk and size of the initially released population (the 'risk-release relationship') are an emerging, high priority field of study [6]. The Brazilian experience shows that there is non-compliance with ballast water management for the Brazilian port authorities [7]. Ever since Brazil signed the Ballast Water Management Convention on 25th January 2005 and adopted its own NORMAM-20 regulations, there has been a decrease in compliance of ballast water regulations with the port authorities. Brazil, with the adoption of national legislation and implementation of an inspection regime require further scientifically validated data for evaluation of its efficacy, besides monitoring and survey campaigns to control the spread of non-native species [8]. It is reported that, more detailed studies are required to assess the reasons for non-compliance and for the most noticeable impacts resulting from them in the waters of Brazil [7]. In European countries, one among the many recommendations that have come up is that all the European Union countries has to ratify the BWM convention which would result in meeting the criteria of the convention and the routine operations of ballast water management systems before the BWM convention enters into force. They are encouraging non-European countries which are bordering European seas to implement common European Union BWM requirements as a pan European application [9]. The major stumbling block appears to be the non-ratification of the convention by countries around the world. Some countries USA, Canada, Australia, Panama, Liberia and the Bahamas has developed unilateral ballast water management legislation which needs to be ratified so that they can monitor the effectiveness of the convention and suggest improvements as technology and compliance advances [10]. After becoming a party for 2004 convention, Malaysian Government has come up with various implications for the stipulated treatment technology as well as for monitoring activities. This would be the major undesirable result which the stakeholders should have to bear when the IMO convention 2004 comes into force [11].

Canada was the first among a few countries to develop a centralized model of control for ballast pollution after the destruction of its marine sanctuaries by bio-invasions. The Canadian law is called the Ballast Water Control and Management (2001) regulations. The law stipulates that every effort by the master and crew has to be performed to minimize the ballast water exchange in the Canadian water or atleast make them harmless before discharging in the Canadian waters through BWE, treatments, retentions and discharge into reception facilities. Within the law distinctions are clearly set for transoceanic ships and non-transoceanic ships when ballast water exchange is done. Transoceanic ships from outside Canada are not encouraged to make ballast water exchange in the mid-oceans. The law directs for a clear ballast water management plan, which should describe the ballast water management process, safety procedures, sediment disposal procedures, design specifications officers in-charge for monitoring and for coordinating BWMP with officials.

3. Systems and Operations in Ballast Water

3.1 Ballast water exchange

The 2004 convention prescribes ballast water exchange as an interim method for prevention of bio invasion till an effective ballast water management plan is in effect for all member states. According to the convention ships have to exchange 95% of the ballast water volume and organisms from the ballast water. The ballast water standards are set based on the ship's age and its capacity. The convention, stipulates that ships shall undertake ballast exchange at 200 nm (nautical mile) from the nearest land and at water depths of 200 m. If it is not possible BWE may be done at 50 nm from the nearest land and at 200 m depth. It also states that during emergency situation, when the prescribed distance and depth measures cannot be maintained, the port states can designate BWE areas with the time required, shipping route and safety requirements kept under consideration. But new suggested route by port states can cause undue delay to the ships further resulting in payment of heavy compensation by vessel owners.

Many studies have proved that the BWE is not a permanent solution to the problem of bio invasion as it can remove only 95% of the organism with one time exchange [12]. Moreover, the organisms settled in sediments may not be removed by single BWE. It is observed that conducting two to three BWE can only help reduce the bio invasion. Ballast water exchange is being gradually phased out depending on the age of the vessel and ballast water capacity. Most of the vessels are in the transition of moving onto ballast water treatment systems from the process of ballast water exchange.

3.2 Ballast water treatment

By 2001, the world maritime community recognized that the BWE alone cannot provide solution to the problem of bio-invasion. There is a need for an alternative method in ballast water management. Researchers found that BWT can be effective in managing ballast water compared to ballast water exchange. The IMO convention 2004 has given guidelines (G8, G9, G10) for the approval of different treatment systems for ballast water management. The ballast water convention does not provide specific requirement for treatment methods to be followed in the BWMS. The treatment designs are reviewed, approved, installed and operated by a Type Approval to be in compliance with the IMO convention for an effective prevention and elimination of invasive species.

Ballast water collected from fresh water, estuary or sea water may contain physicochemical parameters, aquatic organisms and sediments as pollutants. The technologies currently available in inactivating organisms and treating the pollutants are grouped into three categories – Mechanical, Physical and chemical.

It is observed that some of the treatment systems which use chemical biocides or de-oxygenation may require additional treatment prior to water being discharged into the sea. The chemical biocides produce toxic by-products which has so far been land tested and models are developed based on the results. The risk assessment of ballast water treatment systems within the IMO approval procedure is primarily based on exposure from the land based testing and modeling done based on laboratory results.

There are variety of combinations of mechanical, physical and chemical treatment technologies available in the market. Most companies promote combinations for a better result and cost effectiveness. The Article 4 of the convention states that every ship has to exercise control of the transfer of harmful aquatic organisms and pathogens through ship's ballast water and sediments. In order to meet the standard requirements the ships need to conduct ballast water exchange and other ballast water management. This includes treatment systems, all associated control equipment, monitoring and sampling facilities. The systems are required to meet standards of regulation D2 and the conditions established in regulation D3 of the convention. Recommendations regarding the design, installation, performance, testing environmental acceptability and approval of ballast water management facility are provided in the guideline. The ballast water treatment equipment has to undergo Type Approval for operations with active substances and without active substances. The procedure followed are pre-test evaluation of system documentation, Type Approval tests, issuing a Type Approval certificates, and on-board inspections.

With the growing awareness on marine pollution different companies started research and development for the management of ballast water. From the 2004 convention onwards the research in this domain gained new directions and guidelines for the innovation of new ballast water management systems developed in compliance with the D1 and D2 standards. The 2004 convention was ratified by 2016 with a much more effective guidelines and methods for BWMS. With these guidelines the BWT convention came into force by 8th Sept 2017. Majority of the vessels are still in BWE mode and are gradually shifting to the ballast water treatment systems.

The scientific community and stakeholders introduced a number of viable, practical and effective management solutions by 2003 after much deliberation over ballast water management methods. It is found that a single technology will not be suitable for all vessel types and voyage profiles. According to researchers a combination of treatment technology with filtration (primary treatment) followed by biocidal treatments (secondary treatment) based on the vessel type has to be used for a better management of ballast water [13]. According to World Maritime University (Sweden) with other Stakeholders developed many commercially available treatment systems. Among them, 7 of the systems had Type Approval certificate while 20 were in various stages of approval process. Besides the testing protocols the scientific community evaluated the provisions for reception facilities in ports, regulatory, technical and environmental challenges to be in compliance with the convention. They also considered the challenges faced by ship owners in ratifying the convention.

With the increasing Type Approved Systems 30–40 systems are different stages of development, the maritime community considered on sampling of ballast water, monitoring analysis and risk assessment. They observed that still there is dearth in the operational experience for all the available technology in sampling, monitoring and risk assessment of Ballast Water Management Systems. Further, maritime researchers found that ultra violet light used in water treatment can be a better and effective physical treatment method for ballast water. IMO then focused on finding alternatives to on-board treatment systems port based contingency measures, mobile ballast water treatment facilities and treatment boats placed in ports. By 2016 national maritime administrations, ship owners and operators, ship builders and repair yards, test facilities, commercial treatment system manufacturers, research and development communities and financing communities were brought to a global platform for ballast water management.

3.3 Port reception facility

Many ships cannot perform ballast water exchange in the mid-sea due to safety and adverse weather situations. Under these circumstances, they use reception facilities in ports given in the G5 guideline MEPC (Marine Environment Protection committee), 2006. Ships use reception facility for ballast water and sediment management in ports. But it is not mandatory under the convention that the port states should provide this facility for ballast water. The establishment of reception facility requires exorbitant expenses in setting new treatment plants, new pipe connections and more human resources both on board and in the port. Majority port states especially ports in developing countries would advise for ballast water exchange or ballast water treatment as an alternate for port reception facilities.

3.4 Technology approval process

There are many technologies that produce or utilize a substance that has a general or specific action on or against harmful aquatic organisms and pathogens. These substances are called Active Substances according to the Regulation A-1.7. If a ballast water treatment facility uses an Active Substance, then to comply with the Convention it should be approved by IMO in accordance with the 'Procedure for approval of ballast water management systems that make use of Active Substances – G 9' {adopted in MEPC (The Marine Environment Protection Committee) 53rd session}. The following steps has to be followed if a system uses an active substance as per G9 guidelines. It comprises: 1) Initial approval of environmental impact of discharged ballast water {GESAMP- BWWG, Joint group of experts on the scientific aspects of marine environmental protection}. 2) Approval of the system through land-based testing and shipboard trials received by the Flag state. 3) Final approval of environmental impact of discharged ballast water received by GESAMP BWWG. 4) An approval certificate issued by the Flag state. The certification of a system that does not use an active substance should be conducted by skipping the second and the last steps. The testing procedure is outlined in the IMO's Guidelines for Approval of Ballast Water Management Systems (G8 Guidelines). As mentioned, the potential technology should be evaluated during both shore-based testing (6 weeks to 6 months) and ship based testing (6 months). Worldwide, there are few testing facilities for evaluating treatment technologies for ballast water. Since the implementation of the convention, many technologies have been certified [6]. From the time the BWM convention, 2004 and the initial adaptation of the G8 guidelines for approval of ballast water management systems in 2005, substantial number of treatment systems have been developed globally.

3.5 Sampling ballast water

Sampling of ballast water needs to be in compliance with D1 and D2 guidelines. Sampling design given by G9 does not suffice different scenarios of water being ballasted into ships. Samples degrade very quickly and need to be analyzed immediately which depends on the time, place, and holding condition. Moreover, safe limits could not be finalized before sampling process and analysis. More extensive work has to be done towards sampling design and procedure for ballast water. Appropriate physicochemical parameters has to be included in the analysis of ballast water. Inter calibration experiments cannot be compared as there is no uniform sampling methodology currently available for ballast water to prove compliance.

Ballast water management convention has been working on, to evolve scientific sampling methods based on conditions of water and sample sizes. But it was difficult to come to consensus for such complicated issues such as how to sample and analyze ballast water for different organisms. Hence, it has been addressed very loosely in G8 guidelines. Some of the research organizations developed their own standards and sampling methods, with difference in specific details but within the general G8 guidelines. EPA, 2002 has come up with a sampling design. Initially, 1.0 m^3 sample of ballast water was collected to quantify concentration of living organisms larger than 50 micrometer after treatment. Complying with the D2 standard of less than 10 organisms of 50 micrometer size, quantification of the 10 organisms of 50 micrometer size would be easy with 1.0 m^3 ballast samples. Enumeration of the organisms present is represented by the Poisson distribution, and therefore the cumulative or total count is the key test statistic [14]. Further, a chi-square transformation can be utilized to approximate the confidence intervals. Assuming, for organisms $\geq$50 μm, the desired minimum precision in the upper bound of the chi-square statistic should not exceed twice the observed mean (this corresponds to a coefficient of variation of 40%) and count of 6 organisms is required. The volume required to successfully count 6 organisms is dependent on the gross water sample volume, concentration factor, number of sub-samples counted, and the target concentration. For enumeration using subsamples, statistical analysis indicated that 30 m^3 must be sampled to enumerate 10 organisms per m^3, with the desired level of precision. The complexities associated with minimum sample volumes raise additional important issues.

Counting of 10 organisms of 50 micrometer size in such a large volume of water sample is susceptible to error [15]. In none of the studies, count of organisms was made with accurate precision. Samples degrade very quickly and need to be analyzed immediately which depends on the time, place and holding condition. These studies could not finalize the maximum preservation time required before a sample is processed and analyzed [15]. As there is no uniform sampling methodology currently available for ballast water, the biological results obtained from different ships to prove compliance, cannot be compared without inter calibration experiments. Moreover, vessels which show compliance in one port may not be in compliance with another port. In order to get a good representation of the organisms and other chemical parameters of the ballast tank, a most suitable sample access point and frequency of sampling has to be selected accurately. By adopting combination of different sampling equipments a greater range of taxa can be obtained than from any single method. Larger organisms may also be sampled by the use of different collecting methods, such as light traps or baited traps. However, this approach is time-consuming and requires installation of traps prior to sampling which is impossible for control sampling to be in compliance [16].

The ballast water sampling guidelines are mostly used on a trial basis and different organizations develop methodology suitable for their purpose but within the guidelines of the convention. The convention is not able to provide any specific sampling, analysis protocols and legal requirements that can be adopted by any administration. Some member states and ship owners propose to avoid sampling and insist on BWMP, Type approval certificate of BWMS and Ballast water record book for compliance with the convention.

3.6 Analysis methods of ballast water

Different analysis methods available for ballast water samples are DNA method, RNA method, ATP methods, chlorophylla method, oxygen measurement, pulse amplitude modulated fluorometry, flow cytometry, holographic microscopy, visual inspection, and stereo microscope. Presence of DNA and RNA in water can be used as an indication of presence of phytoplanktons in the water. ATP methods are used for detection of viable organisms in water. chlorophylla, oxygen measurement, Pulse amplitude modulated fluorometry detects the presence of phytoplankton in water. DNA, RNA, ATP, chlorophylla, oxygen measurement, PAM, flow camera, holographic microscope methods can be used for analysis of organisms less than 50micrometre size but greater than 10 micrometer size whereas DNA, RNA, ATP, visual inspection, stereomicroscope, flow camera can be used for the detection of organisms greater than or equal to 50 micrometers in minimum dimension [15]. These methods can be used by ports to assess whether it is compliant with the D2 standard of the convention.

3.7 Risk assessment

The convention has outlined guidelines for assessing the risk involved in carrying out ballasting and de-ballasting in relation to granting an exemption in accordance with A-4 of the convention. There are three methods for risk assessment: Environmental matching risk assessment, species' bio-geographical risk assessment, and species specific risk assessment.

Environmental matching risk assessment compares the environmental conditions between locations. Species bio geographical risk assessment compares the overlap of the native and non–indigenous species to evaluate the environmental similarity and to identify high risk invaders. While species specific risk assessment evaluates the distribution and characteristics of identified target species [17].

Environmental matching risk assessment evaluate the salinity, temperature, nutrients and oxygen of the donor and recipient ballast water. The seasonal variations in surface and bottom depths of both the environment are evaluated. If the water is well mixed over the entire year, evaluation of salinity, temperature, nutrient and oxygen depth profiles is not required. If organisms present in the donor regions are tolerant of extreme environmental conditions and can survive in the recipient environment then species specific risk assessment has to be done. Species bio geographical risk assessment compares the distribution of non-indigenous, cryptogenic and harmful native species that presently exist in the donor and recipient ports and biogeographic regions. If the species present in donor port has invaded other biogeographic regions and other related environments, then the organism pose a high risk to the recipient port that has the potential to affect health, ecology and economy of the region [17].

Species specific risk assessment identifies target species which has the potential to impair the environment, human health, property or resources and to survive or complete its life cycle in the recipient port. In this assessment they compare and identify the characteristics of species which has the capability to transfer, survive and reproduce in the new environment. Species specific assessment are done when the donor and the recipient ports are in different biogeographic region. Species specific data with respect to its characteristic behavior in the new environment is very much important for analyzing the risk scenario. More the number of species in the invasive list along with non availability of information on the characteristics of the species more are the chances for risk [17].

Parties may undertake the risk assessment themselves in order to grant exemptions or require the ship-owner or operator to undertake the risk assessment. The recipient port can reject any application for exemption when found not to be in accordance with the guidelines. The exemption has to be renewed every five years from the date it is granted permission. New data and information has to be submitted to show compliance to the exemption.

3.8 Alternative methods

Regulation B-3.7 directs that other methods of ballast water management may also be accepted as alternatives to the ballast water exchange and ballast water treatment, provided that such methods ensure atleast the same level of protection to the environment, human health, property or resources and are approved in principle by IMO's Marine Environment Protection Committee (MEPC). Over the years, new alternative methods to ballast water exchange and ballast water treatment has been introduced into the global shipping community. Some of them are 'no ballast', 'zero discharge', 'ballast free' and 'continuous flow' methods.

In 'No ballast' water eliminates the risk and avoids any ballast water management requirements. This method uses new hull design for the ship or use of solid ballast TEU (Twenty foot equivalent unit) to provide unladen stability and trim without need for ballast water. It avoids cost from fuel and greenhouse gas emissions but has higher hull build costs, operational costs from increasing hull drag and cost incurred from the logistics of handling additional solid ballast TEUs. The stability and buoyancy of the ship when in unloaded condition without ballast water was achieved by widening the ship's beam and moving the displacement volume outward from the centreline by Delft University of Technology (DUT), Netherlands and Det Norse Veritas (DNV), Norway. In the design developed by Daewoo shipbuilding & marine engineering (DSME), Korea the conventional displacement hull retained as the ballast water is replaced by 25 tonnes solid ballast TEU containers. The 'no ballast' water method is applicable to new ro-ro pax, car, containers, livestock ships and other high volume cargo ships.

'Zero discharge' or 'minimal discharge' uses storm ballast water, internal ballast water, and potable water for stability. In storm water, the ship is designed as V-shaped hull which alters the vertical distribution of hull buoyancy causing a deeper draught in unloaded condition. It avoid costs of installing and operating a large BWT systems but need more investment in building the hull. New bulk carriers especially liquid carrier vessels use storm water for ballast water management. 'Internal ballast' concept uses fresh water, which is shifted from one tank to another tank are relatively small to control trim based on the cargo distribution and loading/unloading patterns and are not routinely discharged in ports. Internal ballasting vessels has reduced cargo capability and capacity to make air-draught adjustments. Using this system can avoid the installation and operation of ballast water treatment system. New container ships, ro-ro pax, liners and livestock carriers use internal ballasting. 'Potable water' method use fresh water for filling up the ballast tanks so it can be discharged in any port. By using 'potable water' the ship owners can save cost of installation and space for ballast water treatment system. They usually use modular and compact reverse osmosis systems where the membranes are protected by two or three filtration stages. Depending on the unit size and available power supply these units produce 2–30 tonnes of fresh water per day which is very cost intensive [17].

Continuous flow method is used by longitudinal trunks and ship buoyancy control. In longitudinal trunks, buoyancy trunks replaces ballast tanks to enable continuous flushing without pumping. The longitudinal trunk, flush out the water as much as possible within 1-2 hrs at normal unladen voyage speed with minimal retention of sediment and the organisms will be carried only for short distances. It reduces the cost of installing and operating a large ballast water treatment system but requires high investment for building the ship and valve servicing.

Ship buoyancy control method commonly uses multiple below waterline valves. They convert each ballast water tank into a free flooding buoyancy compartment for continuous flushing without pumping. It has benefits of not installing and operating a large ballast water systems but has to spend on installing valves, control system, valve servicing, coating and cleaning cost to maintain ship safety and bio-fouling. Enhanced ballast water exchange concepts are used in AUBAFLOW, Loop ballast exchange and Dyna ballast where AUBAFLOW and Loop ballast exchange use enhanced blue water BWE by flushing without using pumps for transoceanic voyages and Dyna ballast uses a specialized aerator- educators to all ballast water tanks. Since there are no ballast water treatment available in the continuous flow method, sediment and organisms in the low flow zone can cause impedance to the D2 standard. As compared to normal ballast water exchange, ballast pump servicing costs, fuel consumption, and greenhouse gas emissions will also increase.

4. Conclusion

The 2004 convention on ballast water management is found to be inadequate in providing a solution for a complete elimination of bio-invasion and the already invaded organisms. It provides BWE as a solution to the ballast water management with a specified distance and depth at which it can be performed. It could not give directions on how to perform BWE at the designated site during an emergency situation. It also do not give explanation on how to select the designated area for each port states. The convention fails to address the delay in issuing port clearance for some ships due to difference in sampling facilities available in different countries. Amendments has been brought to the 2004 convention over the years for development of ballast water management plan, ballast water exchange, approval of methods used for ballast water treatments, and control of ballast water and sediments. By 2018 an elaborate regulation and guidelines were provided for covering every aspect of ballast water management.

The member states should share technology to developing countries in establishing sampling and testing laboratories. The monitoring of the system could not be accomplished due to lack of proper training to the port officers and equipment transfer. Proper training of Port officers or each state head for maritime management in sampling, monitoring, analysis and other ballast water management systems should be conducted by IMO worldwide.

Region specific sampling methods and analysis has to be developed for each country based on the occurrence of bio-invasion specific to the region. Region specific elaborate research and data has to be generated to understand the characteristics of different target species in evaluating the risk assessment. New guidelines has to be formulated and implemented to manage the already established invasive species in each country.

The D2 standards should be changed from size specific standard to 'no organism' to be present in ballast water for discharge. Ships gradually transiting from ballast water exchange to ballast water treatment can find new combination of ballast water treatments for achieving the 'no organism' standard. Ships which are newly constructed should invest in new alternate designs of 'zero ballast' or 'no ballast' or 'continuous flow method' so that no organisms are harmed in the quest for achieving the regulations and guidelines stipulated by the convention. The zero ballast method can bring in balance and sustainability in the marine ecosystem.

Acknowledgements

I would like to thank Cochin University of Science and Technology, Kochi, Kerala, India for providing an opportunity to work on ballast water.

Abbreviations

BWM	Ballast water management
BWT	Ballast water treatment
BWMP	Ballast water management plan
BWE	Ballast water exchange
BWMC	Ballast water management convention
BWMS	Ballast water management systems
IMO	International Maritime Organization
MEPC	Marine Environment Protection Committee
UNCLOS	United convention on the law of the sea
UNDP	United Nations development program
GEF	Global environment facility
EPA	Environment protection agency
TEU	Twenty foot equivalent unit
DNA	Deoxyribo nucleic acid
RNA	Ribo nucleic acid
ATP	Adenosine triphosphate
MARPOL	International Convention for the Prevention of Pollution from Ships
SOLAS	Safety of life at sea
ISM	International safety management

Author details

Eleyadath Lakshmi, Machinchery Priya and Velayudhan Sivanandan Achari*
School of Environmental Studies, Cochin University of Science and Technology, Kochi, Kerala, India

*Address all correspondence to: vsachari@cusat.ac.in

References

[1] Gollasch S. Lenz J. Dammer M. Andres H.G. 2000. Survival of Tropical Ballast Water organisms during a Cruise from the Indian Ocean to the North Sea. Journal of Plankton Research. 22. 923-937.

[2] Castro MCT. 2012. Implementation of Ballast Water Management Convention, 2004 – Background Information on the Subject and Enforcement Procedures. The United Nations–Nippon Foundation Fellowship Programme. Division of Ocean Affairs and The Law of the Sea. New York, United Nations.

[3] IMO (International Maritime Organization). 1997a. Guidelines for the control and management of ships' ballast water to minimize the transfer of harmful aquatic organisms and pathogens. IMO Assembly Resolution A.868.20.

[4] IMO (International Maritime Organization). 1997b. Guidelines for the Control and Management of Ships' Ballast Water to Minimize the Transfer of Harmful Aquatic Organisms and Pathogens, Resolution A. 1997.868 (20). London, UK.

[5] Oliveira UC. 2008. The Role of the Brazilian Ports in the Improvement of the National Ballast Water Management Program According the Provisions of the International Ballast Water Convention. The United Nations-Nippon Foundation Fellowship Programme 2007-2008. New York

[6] Tsolaki E. Diamadopoulos E. 2009. Technologies for Ballast Water Treatment: a review. Wiley Interscience.

[7] Castro MCT. Jason MHS. Cecilia FP. Timothy WF. 2017. Ten years of Brazilian ballast water management. Journal of Sea Research. 133. 36-42.

[8] Lehtiniemi M. Outinen K. 2017. Literature review for the indicative ballast water analysis methods. Technical report. Finnish Environment Institute.

[9] David M. Pirelli F. Petri A. Gollasch S. 2016. Compliance monitoring and enforcement measures and decision support systems for implementation of the BWM Convention in Adriatic, including reviews, models and test results of compliance control measures according to the BWM Convention. Final report. BALMAS project. 44. (Figure 2 reprinted from the article by Ferrate treatment technologies: A leader in water treatment technologies. www. link. springer.com)

[10] Anstey F. 2008. Ballast water management: A time for action. Fisheries and Marine Institute of Memorial University of Newfoundland. Canada.

[11] Kaur R C. 2014. Progress Challenges and the Way Forward on Ballast Water Management in Malaysia, Indian Journal of Geo-Marine Sciences. 43.11.

[12] Vijayan S. 2014. Indian Law on Control of Vessel Sourced Pollution in Maritime Ports [thesis]. School of Legal Studies, Cochin University of Science and Technology. Cochin.

[13] Lakshmi E. Priya M.Sivanandan Achari V. 2021. An overview on the treatment of ballast water in ships. Ocean and coastal management.199.

[14] Lemieux EJJ. Grant T. Weir S. Robbins S. Riley W. Hyland L. Davis T. Donowick B. Stringham W.J, Kinee B. Brown and Everette R. 2008. Private Environmental Technology Verification (ETV) Test Report of the Seve Trent DeNora Bal Pure ballast water treatment system. NRL letter report 6130/60978. Washington, D.C.

[15] Gollasch S. David M. 2010. Recommendations How to Take a

Representative Ballast Water Sample, Emerging Ballast Water Management Systems. IMO-WMU Research and Development Forum. Malmö, Sweden.

[16] Gollasch S. Stehouwer PP. David M. 2012. BWO Technical outline and requirements for organism detection systems for establishing compliance enforcement. Summary of Final Report, Interreg IVB Project 5.

[17] Glo Ballast Monograph Series. 2011. Establishing Equivalency in the Performance Testing and Compliance Monitoring of Emerging Alternative Ballast Water Management Systems. A Technical Review. IMO, UK.

Section 2

Modern Ship Management

Chapter 3

Private and Public Management of the Real Estate Registry in Chile

Santiago Zárate González

Abstract

The present work seeks to reflect on the best real estate registry management system in Chile, from a private one, such as the one that exists today, or a public one; examine their strengths and weaknesses; and to what extent both are convenient to use technology. Chilean legislation will be reviewed for this under a historical and also dogmatic perspective.

Keywords: management, registration, real estate, Chile

1. Introduction

Since 1855, Chile has had a real estate registry system, which was designed for a world different from the one in which we live today.

Our Civil Code, promulgated in 1855, came into force on January 1, 1857, with the exception of its article 695, which gave the President of the Republic the mission of generating a regulation that would regulate in detail the functioning of the system, from the figure of a Conservator of real estate, derived from the French mortgage legislation (*Conservateur des Hypothèques*), to how to practice the entries of registration, modification, and cancelation of the rights with access to the registry.

The system did not enter into force until the aforementioned regulations were issued, which occurred on June 24, 1857, and after some setbacks, the system came into operation on January 1, 1859. On that date, the rules of the Civil Code on the real estate registry system finally came into force.

After the aforementioned date (1859), some formal modifications were made, without going into the substance of its rules. Such modifications referred, for example, to the removal of the rules on the appointment and functions of the real estate conservators, from the 1857 regulations to the Law on the Organization and Powers of the Courts of Justice, enacted in 1875.

After this legislative milestone, the State of Chile amended the law in question, giving rise to another law that is still known today as the Organic Code of Courts, which replaced the law of 1875, introducing the generic denomination of auxiliaries of the administration of justice, which included notaries and real estate conservators, by the way. Thereafter, there have been minor amendments to legal bodies related to taxation, which have not reformed the system in its essence.

Already in the twenty-first century, and due to the technologicalization of the notarial and registry activity, in some minor aspects, a digital registration system was created for some types of companies, which exists in parallel to the paper or physical system, created at that time by mandate of article 5 of the Code of Commerce, dictated in 1865.

The idea, very different from that of the real estate registry, was to give strength to the creation of companies without the bureaucratic procedures that had to be complied with by those who chose to incorporate a commercial company. In the same way, the government of that time wanted to avoid the costs involved in incorporating a commercial company, such as the minutes of a lawyer, the public deed of incorporation, the extract, its publication in the Official Gazette of the Republic, and its registration in a special registry, still kept by the Real Estate Registry, all within the fatal period of 60 days.

However, the entry made in the commercial registry did not create a personal right different from the one created by virtue of the partnership agreement; rather, such entry in the commercial registry has always and to date only served the purpose of formal, and not material, publicity.

The new system shortened incorporation times from 60 days to one day. That is to say, in one day the future partners can have a company incorporated free of charge. Registration is done automatically in a special commercial registry which is kept by the Ministry of Economy, and not by a legal professional, as is the case of the parallel and older commercial registry which is kept by the Real Estate Registry.

The fact of creating companies in one day produces many benefits to those who opt for this type of human associations. By way of example, improvements in the costs of incorporation of the companies which are close to "0"-, of operation, of certifications, of procedures before tax and accounting services, etc.[1]

However, other problems arise that are related, among other things, to the control of such entities in relation to the delay in the initiation of activities, or the making of the first Monthly Provisional Payments (PPM), or the making of the first sales or services, and others that refer to the fact that most of the data required in the form that is filled out online, are nothing more than declarations that may well be dissociated with reality; for example, there is the issue of the domicile in which the company will have its head office, so that government agencies, such as the Chilean Internal Revenue Service (SII), have many problems to control these incorporators, which are called for that reason "taxpayers of difficult or no control," which already existed before, but now, have been accentuated.[2]

In 2002, Law 19.799[3] was enacted, creating the so-called "Advanced Electronic Signature" (FEA), which was intended to put an end to the endless lines of users of public services, the judiciary, notaries, and real estate registries. By means of a simple certification and validation mechanism, it was possible to carry out procedures that previously required the presence of the requesters.

In 2018,[4] the government presented a bill in which it intended to introduce reforms to the real estate registry system, mixing notaries and judicial archivists in the process, a matter that clearly turned out to be inappropriate and ineffective in practice. Despite the fact that the project is advancing at a dinosaur pace, the changes will come whether notaries, conservators, and archivists want them to or not.

Most of the proposed modifications are formal so that the substance of the real estate registration system was not altered, all of which has resulted in a loss of

---

[1] See https://www.economia.gob.cl/wp-content/uploads/2016/01/Bolet%C3%ADn-RES-2015.pdf

[2] See https://www.sii.cl/preguntas_frecuentes/iva/001_030_3496.htm in which letter B refers to taxpayers (corporations) that cannot be effectively audited. Also in Alcalde Silva et al. ([1], pp. 248-249).

[3] See https://www.bcn.cl/leychile/navegar?idNorma=196640

[4] See https://www.alejandrobarros.com/wp-content/uploads/2018/11/Proyecto-de-Ley-Notarios-y-Conservadores.pdf The bill is currently in its second legislative stage in the Chilean National Congress.

resources for the State after a long parliamentary discussion, which has been slowed down by the social outbreak of October 2019 and the pandemic that is still not completely over.

In what is relevant to this work, and which is noteworthy, the project proposes a series of technological modifications, including online platforms and registries in charge of state agencies, such as the Civil Registry and Identification of Chile, a public service that is far from being efficient.

What is peculiar is that the legislator intends to hand over some matters to an inefficient and technologically permeable public body, while maintaining the position of Real Estate Conservator. A parallel system is created that has no genuine meaning in itself.

What then is the sense of applying technology to improve the important aspects in the processing of registrations, entrusting the keeping of records to a legal professional such as the Real Estate Registrar, while incorporating a partner called the Civil Registry that will keep parallel records containing the same documents as the registrar?

In this paper, we will therefore try to answer this question and others that will surely arise along the way, in order to establish which is, in our opinion, the best system to operate the real estate registry system in our country.

2. Change from a paper format to a virtual or digital format

The paper format in the records which is kept by the Real Estate Registry has been in force for a little more than 160 years since it was created by the Civil Code of 1855, and later developed in the Regulations of 1857.

This system has undoubtedly served all those who today can say that they are owners of real estate property, or of rights in rem constituted therein. Banks and other financial institutions have achieved a level of certainty and legal security in real estate and credit transactions that no other system could have given them.[5]

This does not admit much discussion, despite the comments and opinions of the notaries and conservators' union sector that are reluctant to make changes to the system. It is true that the 2018 project does not make serious incursions in substantive aspects of the system, but it announces in a brutal way that the will exists in the congressmen to evolve, even if it means going against the current. In this regard, the silence of the system's financial operators, who may not benefit much from the changes, especially with regard to the costs and securities currently offered by the system, is striking.

If the reform intends that the real estate registry transits from the paper that has given news of legal certainty, why is it about modifying rules that have nothing to do with the substance of the system?

Except for what refers to the keeping of books by the real folio mechanism, the truth is that the 2018 reform does not innovate in other central aspects of the system, such as the keeping of partial records. That is not understanding how not only our system works, but all the systems in the world. But what kind of world is that?

Based on what was raised by the National Economic Prosecutor's Office in 2018,[6] and the reasons underlying the project of the same year (to which we have alluded previously), it turns out that the vision of the prosecutorial body of activities

---

[5] See Mohor Albornoz ([2], pp. 8-9). Also in Gutiérrez González [3].

[6] See https://www.fne.gob.cl/wp-content/uploads/2018/07/Informe-Final-optimizado.pdf which refers to the presence and operation of the notary's office in the metropolitan region of Santiago and Valparaíso.

contrary to the free market and the support for it from the government translated into the aforementioned project are the two sides of the same coin—the influence of the economic rules of a system whose fundamentals differ from ours. In effect, it is a matter of applying economic rules typical of Anglo-Saxon countries to a reality that does not comply with continental legal principles such as those that govern systems like the Chilean one.

The lightness with which issues such as notarial and registry public faiths are treated seems to us to be an attack against legal certainty. A market-based model, which seems to be the panacea of the moment, is an issue that we do not share, precisely because of the need for activities such as notarial and registry activities to be based on principles of probity and transparency that the market does not properly ensure. Business ethics is far from being accepted in English-speaking countries.

Now, what does this have to do with replacing the paper format with a virtual one? The truth is that not much, *a priori*.

Basically, the fact that registrations or public deeds are recorded in virtual instruments is not new. As we have said, it already exists in laws such as the one creating the FEA, or the electronic commercial registry for companies.

What is there than in the paper that does not exist in electronic or virtual? For one thing, legal certainty, both are dynamic and static.

In a world in which everything is electronic or digital, the paper format offers security in terms of the fact that everything that is recorded in the physical record is covered by a halo of legitimacy that the electronic format unfortunately does not yet have.

Let us take as an example, the real estate financing carried out through online operations, which today reaches aspects of the real estate registry, through the *Blockchain* mechanism, and *Tokenization* [4].

In the first case, a Blockchain can be defined as a "shared digital ledger comprising a list of blocks connected and stored in a distributed, decentralized and cryptographically protected network, serving as an irreversible and incorruptible information repository" ([5], p. 63).

According to Castiñeira, it is "a technology that allows the transfer of securely encrypted digital data" [6]. This transfer, "continues the author," does not require a centralized intermediary to identify and certify the information, but is configured through nodes (nodes, if you will) independent of each other, which register and validate the transfers without the need for prior knowledge or to generate a situation of trust between them." [6]

In the second case, tokenization comes to complement the chain in the sense of "abstractly representing a value through the blockchain" ([5], p. 61).

It is assumed that, through these mechanisms, both transactions and registration are transferred to the electronic or virtual environment, promising higher security standards than those provided by the paper format.

However, and as we have seen in other opportunities, these modern and digital mechanisms generate conflicts that, in general, are related to their lack of regulation by nations; the need for greater acceptance by legal systems; fiscal consequences arising from the discussion about their nature (which in any case does not offer major difficulties), and the application therefore of exaggerated taxes, thus causing a diminished legal security (inflation and high costs, lack of transparency and use of cryptocurrency, as a basis for the generation of profits) [6].

In Chile, these technologies (like so many others before) have not managed to be introduced in areas other than the financial one, for the same reasons mentioned by Castiñeira, which may vary by the will of the legislator, which can give a boost to the acceptance of these mechanisms. Slowly goes a long way, says the adage, and to that extent, there are many possibilities that both the *Blockchain* and the *token* permeate the domestic legal order, mainly from its acceptance and development at the legislative level, applying it to various facets of national activities, starting with the economy and finance, thereby crossing the protective membrane of the law, *lato sensu*.

In the words of Pacheco Jiménez, there is a kind of "regulatory uncertainty," which causes distrust in the nations where it is applied ([5], p. 72), stemming from cases of scams or frauds with cryptocurrencies in the USA. In fact, the Texas Securities and Exchange Board issued an Emergency Cease and Desist Order against Forex Birds and PEK Universe, two financial intermediaries that allegedly defrauded some people by operating from abroad, which obviously makes it very difficult to control both companies and transactions.[7] In Chile, without going any further, there have also been crypto scams.[8]

In this way, it seems that the system is not so secure yet, thinking especially about what refers to cryptocurrencies and the mechanisms associated with them (*Blockchain* and Tokenization), which makes you think about what would happen if they were applied to the real estate registry field.

We refer to public faith as a legal principle protecting the real estate registry. In countries such as Spain, public faith seeks to protect the acquisitions made by third party purchasers once they have been entered into the registry.

According to Gordillo Cañas, plainly and simply, the public faith of the registry "means as much as objective reliability of the Land Registry: everyone can trust (*fides publica*) in what the Registry publishes" ([7], p. 510). For Diez-Picazo "There is public faith in the registry to the extent that third parties can place their trust in what the Registry publishes [...]" ([8], p. 450). Finally, García García points out that it is a (mortgage) principle "[...] by virtue of which the third party who acquires on the basis of the dispositive legitimacy of a registrant is maintained in the acquisition a non domino that he makes, once he has registered his right, with the other requirements demanded by the Law" ([9], p. 227).

The trust in the paper-based registry system is what legitimizes it so that since there is no security in the use of electronic mechanisms and devices for making the most important entries, the truth is that the possibility of applying them to more substantive issues related to the real estate registry, other than financing, still feels distant.

Our country has begun a journey toward technologies related to formal aspects of the various activities related to the registry, demonstrating in practice a facilitation of many procedures that were handed over to notaries and conservators, such as affidavits or certifications that can easily be done online, or that have no importance in terms of their effects.

The AEF law and the reforms proposed in 2018 to the notary and registry system, as we have pointed out, take the path of incorporating technological tools that will make them more dynamic. However, it is very likely that no substantive or substantive issues involving the weakening of public faith will be affected, which is, ultimately, the most important guiding principle applicable in the case of Chile, and what must be avoided. Consequently, we believe that the paper format will continue to offer better guarantees of security in both notarial and registry activities, at least as far as substantive aspects are concerned. The formal aspects, although important, allow the use of modern technological tools that facilitate a large part of the formalities that until now were done in person.

It is not a matter of preventing the application of these tools to the execution of registrations and public deeds, *per se*, but of offering to those who participate

[7] At https://www.ssb.texas.gov/sites/default/files/files/news/ENF_20_CDO_1820.pdf

[8] At https://es.beincrypto.com/recopilacion-estafas-bitcoin-btc-criptomonedas-chile/

in the different legal acts and contracts, a high level of confidence that what they have agreed upon will be fulfilled or will produce the effects sought by them. This is what we mean by substance and form, two sides of a coin that moves forward.

3. Private or public management of notarial and registry activities

It is clear that based on the abovementioned, it is relevant to refer to the management of the system in terms of the powers that the law has conferred on notaries, and more specifically, on real estate conservators.

The government's proposal of 2018 moves forward in some formal ways that involves the dynamization of the system, without touching, however, aspects referred, as we have pointed out, to the substance or substance of the system. This, without losing sight of the fact that we do not share what refers to the existence of repositories or files in which the same information is kept by different subjects, and that could imply problems of trust in the information contained in the real estate registry.

The existence of public services that reduce, at least in theory, the costs and time associated with the delivery of the information contained in the registry is not an obstacle to good registry management. That is to say, if the idea is that users pay less for some certifications and copies, there is nothing to prevent the conservator himself from issuing them efficiently. There is no need to hand over this function to a State service that is far from being efficient.

We believe that the conservator can grant this kind of documents and, therefore, is a sufficient guarantee of trust in the delivery of the information contained in the different books that compose the registry. Having the quality of minister of faith, the information becomes authentic or trustworthy, so that it surrounds the issued document with legitimacy.

The interesting thing about the 2018 project is that these actions can be carried out by the conservators through websites created for this purpose, and in which you can find at an adequate cost (which could be reviewed without a doubt by the legislator, given the ease of obtaining them because they are electronic PDF), all those registry actions that are related to the registry, such as copies of registrations, domain certifications, or of the limitations to it, guarantees that affect the real estate, copies of plans, and other existing documents in their files.

Regarding the elaboration of registrations, they can also be subject to electronic applications, which are done nowadays by some conservators in Chile. The registry, to that extent, seems not to be affected; however, the same does not happen with the notary's work, which is prior to the authentic title entering the registry through the registration entry.

Indeed, when an authentic instrument enters the registry for registration, it must be reviewed by the registrar, who carries out a true control of the legality of the title. This is because if the title presented contains any defect that makes it legally inadmissible for registration, this defect will affect the registry in terms of legitimacy so that the system of qualification of authentic instruments that enter the registry must be reinforced to protect third parties operating in it.

We are about to continue with the development of digital platforms in which the registry information can be consulted without the need to pay for it, which seems to be one of the formal objectives of the projects that have wanted to reform the system.

And what, in our opinion, should be the way to avoid conflicts with the public faith of the registry and the legitimacy of the registry acts?

Firstly, to understand that "digital" is not something negative or detrimental to people, both for registrars and for system operators. Digital platforms should be thought of as tools that facilitate the economic and social development of a country, in the broad sense of the term. Therefore, we cannot think that something that facilitates an activity should be understood as a setback.

Secondly, an efficient system must balance legal principles with the reality of modern real estate and credit traffic. Undoubtedly, credit is the engine that drives real estate circulation. The creation of added value offered by the construction industry is evident and necessary in a stage of economic growth that positively affects social health indicators (although it is not the panacea of social emergence and mobility).

The real estate market moves at great speed, and that speed is met with a slow State and subject to a bureaucratic legality that slows down the processes, causing situations of corruption that should not exist in ethical terms. This slowness even appears as something that is added to the cost structure of the projects, thinking precisely about the speed of return of investments, especially those involving large sums of money. In this sense, there is talk of mega real estate projects.

Within this tangle of administrative obstacles, there is that related to the registry activity, since the process of preparing the proper entries for the transfer of ownership, or the constitution of encumbrances, preferably mortgages, is slow due to the advance of technology.

Creating a secure digital system for the preparation of registry entries is necessary today, but not for economic reasons, which may be very reasonable, but for legal reasons of certainty and security.

Therefore, rather than encouraging the development of tools such as *Blockchain* and *Token*, used for the purpose of wealth generation, we should focus our efforts on the creation of secure systems against, for example, attacks by *hackers* and other cybernauts. Therefore, applying this knowledge and keeping it away from greed seem to be the best way to go in the registry activity.

Let us imagine a system based on principles that make the activity secure, and inevitably, we must recognize that the paper format and the vaults in which the registry information is kept and guarded, allowing us to sustain confidence in the system.

Thirdly, changing the management of the registry activity in charge of a legal professional, invested by the State, and handing it over to a State organ, does not solve the problems of slow administrative procedures, because the problem comes from above, from the legislator, who sees in an excessive amount of operations, the answer to the legal certainty that must exist in the traffic. That is to say, the State and its organs direct their efforts to the creation of unnecessary formalities, with the erroneous idea that this protects the system.

Therefore, when the legislator presents his reform projects, he does not intend to solve technical problems of the system, which are at the base of its essential or substantive structure, but rather his efforts are focused on formal aspects that any public policy coming from the government of the day can easily change without the Parliament necessarily having to participate.

Since its establishment in the second half of the nineteenth century, the management of the registry activity has been entrusted to persons trained in law (preferably lawyers), under the protection of figures such as the notary, which is prior to that of the real estate conservator, and which comes from the Spanish legislation applied in Chile until well into the same century [10].

The notary was a very broad figure used in the pre- and post-Independence period in the American nations, to whom the Spanish regulations conferred a preponderant role in the celebration of legal acts with permanent effects, so that its existence was linked to the public trust or public faith; without prejudice that there were also for judicial work.

In fact, once the offices of conservators were created in 1857, they were slow to be established mainly due to practical issues—low stipends and very large territories.

The idea that the custodian of the most important economic rights in a society, such as property, was given to a legal professional who did not belong to the judiciary and was not remunerated by the State, which denotes a degree of civilization and innovation that was very relevant for the time.

The first Chilean registry was organized on the basis of the already existing office of mortgages, censuses, and ships, created by an unnumbered law of 1845, whose Article 24 gave the power to the President of the Republic to issue regulations, which were promulgated on May 20, 1848. It was a real folio registry kept by an official called "tenedor del registro," who followed the Spanish tradition of being a legal professional whose remuneration came from fees fixed by law or by decree. This position was assimilated to that of the notary of private acts, as mentioned above [11].

This first registry follows, in turn, the form and content of the norms on the office of mortgages and censuses that was created by King Charles III of Spain, by Royal Pragmatic on January 31, 1768, and whose norms were later included in the Novísima Recopilación de Leyes de España (New Compilation of Laws of Spain) of 1805.

After 1848, the system evolved toward more modern models existing at the time in Europe, mainly in German-speaking countries. The promoter and ideologist of these models were undoubtedly President Manuel Montt, who, supported by Andrés Bello and José Alejo Valenzuela, gave shape to a hybrid system that mixed the best of the models, turning our registry into a robust and secure system in terms of legitimacy and public faith, giving proof of its strength, especially to financial operators throughout these 160 years of existence.

The paper format had much to do with the process and success, without prejudice to the chaos in bookkeeping that existed in the mid-nineteenth century, and that the new system corrected in a remarkable way.

Therefore, we believe that the private management of the real estate registry in Chile has obtained excellent results both in financial terms and in terms of legal security, following the Spanish tradition of notaries, as custodians of the public faith or trust.

4. Conclusions

Whether in paper or digital format, any real estate registry system must prioritize public faith and legitimacy as essential postulates, which will give it a better performance in terms of certainty and legal security for both credit and real estate traffic.

Telematic systems offer a certain level of dynamism to the registries, in general, which decompresses in a certain way all that maelstrom of administrative procedures, easy to replace.

However, there are still cracks in these new technologies that, although they work in terms of speed in transactions, do not offer security in legal terms, since the information contained in the registries could be circumvented by *hackers* or cybernauts, whose actions would jeopardize the public faith of the registry.

It is not that new technologies do not offer the dynamism that new times require to give a great boost to the economy, especially in times of pandemic. Perhaps the focus should not be on transactions, which are obviously of great importance, but rather on the efficiency of registry acts in technical and substantive terms that generate protection and not weakness. Therefore, perhaps extracting the best of these technologies is the way to be able to apply them. As they are far from money, they would not be the target of frauds or swindles, as associated crimes.

The management of registries by natural persons and legal professionals, with good training in the field, for example, should not be handed over to public bodies, not only because they have proven to be inefficient in the keeping of the so-called factual registries, such as births and deaths; therefore, handing over to them the keeping of legal registries is not a good idea for us.

It is not a question of closing the door to technologies that support the work of the registry offices, since in most cases, these are formal matters that can be perfectly replaced by telematic or digital mechanisms, as in our country has begun to be applied by the law of advanced electronic signature, as we explained above.

The elaboration of registrations, annotations, and other entries made by the real estate conservator can be replaced by digital mechanisms that streamline the work of their offices, turning a system that at times seems slow and old into a speedy one.

Replacing the paper format in this sense does not imply a real affectation of the guiding principles of the activity, and the legislator should emphasize the protection of the contracting third parties, the registered holder, by the way, and of all those who operate with the system, as is the case with the banks and the State.

Like any human work, both the system of digital record-keeping and the processing of the information contained in them must go through the review of the control mechanisms of the activity, which today are minimal and very biased in our country.

Therefore, it is not that the use of technologies is a bad idea in itself, as some would have us think, but to improve the mechanisms of control and management of information in terms of security, both cybernetic (avoiding external attacks) and legal (format). A system that goes in that direction seems to us to be more appropriate to the times and much cheaper for those who give importance to that aspect alone.

Author details

Santiago Zárate González[1,2]

1 Universidad Central de Chile, Chile

2 Pontificia Universidad Católica de Valparaíso, Chile

*Address all correspondence to: santiago.zarate@ucentral.cl

References

[1] Alcalde Silva J. et al. Algunas observaciones preliminares respecto del proyecto de ley que modifica el sistema registral y notarial en sus aspectos orgánicos y funcionales. In: RChDP UDP 33. Santiago: Universidad Diego Portales (Fundación Fernando Fueyo Laneri); 2019. pp. 243-297

[2] Mohor Albornoz E. Organización registral chilena y los desafíos ante las aplicaciones tecnológicas. In: Paper at Congreso CINDER 2014. Santiago: Revista Fojas; Available from: http://fojas.conservadores.cl/wp-content/uploads/2015/05/Ponencia_EMohor_v2.pdf

[3] Gutiérrez González N. Sobre la reforma al sistema registral inmobiliario. Santiago: Revista Fojas; Available from: http://fojas.conservadores.cl/articulos/sobre-la-reforma-del-sistema-registral-inmobiliario

[4] Criado Enguix J. Blockchain: Cryptocurrencies and tokenization of real estate assets. Efectos en el ámbito registral. In: REDS 16. Madrid: Dykinson; 2020. pp. 253-277

[5] Jiménez P, Nieves M. De la tecnología del Blockchain a la economía del Token. In: RD PUCP 83. Lima: Sistema de Bibliotecas PUCP; 2019. pp. 61-87

[6] Castiñeira Carnicero R. Blockchain y contratación electrónica: ¿regulación versus regulador?. Available from: www.regispro.es: https://regispro.es/rafael-castineira-carnicero-blockchain-y-contratacion-electronica-regulacion-versus-regulador/

[7] Gordillo Cañas A. El principio de fe pública registral (I). In: ADC BOE 2. Madrid: Ministerio de Justicia/Boletín Oficial del Estado, BOE; 2006. pp. 509-655

[8] Díez-Picazo y Ponce de León L. Fundamentos del Derecho Civil patrimonial, III, Madrid: Thomson Reuters Civitas; 1995

[9] García García M. Derecho registral o inmobiliario, II. Madrid: Tirant lo Blanch; 1993

[10] Bravo Lira B. La institución notarial en Chile. Notas sobre su naturaleza y configuración jurídica. In: RD PUCV 2. Valparaíso: Pontificia Universidad Católica de Valparaíso; 1978. pp. 63-72

[11] Zárate GS. El primer registro chileno sobre hipotecas, censos y naves regulado en el reglamento de 20 de mayo de 1848. In: REHJ 41. Valparaíso: Escuela de Derecho Pontificia Universidad Católica de Valparaíso; 2019. pp. 399-413

Chapter 4

Management of Nutrient-Rich Wastes and Wastewaters on Board of Ships

Céline Vaneeckhaute

Abstract

Ship-generated nutrient-rich waste sources, including food waste and sewage water, contribute to eutrophication and deoxygenation of marine ecosystems. This chapter aims to discuss the characteristics of these waste and wastewater sources, review current ship-generated organic waste and wastewater regulations, inventory conventional management and treatment practices, and identify future perspectives for more sustainable nutrient-rich waste and wastewater management on board of ships. According to regulations, untreated food waste and sewage can generally be discharged into the open sea at more than 12 nautical miles from the nearest land, hence this is currently a common practice. However, special restrictions apply in special designated areas such as the Baltic Sea, where food waste must be comminuted/grounded and nutrients need to be removed from the sewage prior to discharge at 12 nautical miles from the nearest land. Current research looks at the valorisation of these waste and wastewater sources through anaerobic digestion, composting and/or nutrient recovery.

Keywords: food waste, management, maritime, nutrients, organics, sewage, ship, valorisation

1. Introduction

Nutrients such as nitrogen and phosphorus can cause devastating impacts on the aquatic environment. Although nutrient pollution in rivers and lakes has been widely studied and measures have been put in place to reduce nutrient discharge in these environments [1, 2], much less attention has been paid to nutrient pollution in marine ecosystems. Nutrients can enter seas and oceans through discharge of wastewater and food waste generated either at land or on board of ships, aquaculture practices or fertilizer run-off from agricultural land [3]. Such excessive nutrient discharges into the marine environment, combined with ocean warming due to climate change has resulted in ocean deoxygenation [3, 4]. Globally, the oxygen content of the ocean has decreased by around 2% since the middle of the 20th century [4]. This number is expected to further decrease by 3 to 4% by the year 2100 under a business-as-usual scenario [4]. Much of the oxygen loss occurs in the upper 1000 m where species richness and abundance is the highest [4, 5]. Sensitive species are as such replaced by more tolerant and resilient species, with a decrease of biological diversity as a result [3]. Ocean warming-induced deoxygenation is driven by the fact that warmer ocean water caused by climate change holds less oxygen and is more buoyant than cooler water [5]. This leads to reduced mixing of oxygenated water near the surface with deeper waters, the latter naturally containing less oxygen [5]. Warmer water also raises the oxygen demand from living organisms. As a result, less oxygen is available for marine life [3, 4]. Nutrient-induced deoxygenation is caused by the abundance of nutrients that induce eutrophication, a phenomenon often found in coastal waters [6]. Eutrophication is characterized by the production of harmful algal blooms and may result in oxygen depletion of the water body after the bacterial degradation of the algae [6]. Moreover, certain species of algae produce biotoxins, which are natural poisons that can be transferred through the food web, potentially harming higher-order consumers such as marine mammals and humans [6, 7]. If human-accelerated eutrophication is not reversed, the entire coastal ecosystem may ultimately be devastated.

The significant growth of the maritime transportation sector over the last decade has accelerated the devastating environmental impact on marine ecosystems [8, 9]. The cruise ship industry now transports about 22 million people annually around the world [10]. Moreover, marine policy stimulates transport of cargo by sea, thereby also further increasing the number of people (staff) traveling over sea [10, 11]. This increase in maritime transportation comes with a global increasing amount of ship-generated waste. Nutrient-rich waste on board of ships includes sewage (gray and black wastewater), as well as food waste [9]. Despite the increasing knowledge and concern of the environmental impact of nutrient discharges into the marine environment, these organic waste sources are often still dumped into the open sea without treatment [12]. As an example, ship-generated nutrient discharge into the Baltic Sea has been estimated at 269 tons of nitrogen and 256 tons of phosphorus in the year 2000 [12]. Since ship-generated waste pollution is one of the main concerns of this area, it has been declared as a Particularly Sensitive Sea Area by the International Maritime Organization (IMO) [13]. This means that the area now requires special protection through legislation and actions because of its socioeconomic and scientific importance [13, 14].

This chapter discusses the source of nutrient-rich wastes and wastewaters produced on board of ships, current ship-generated organic waste and wastewater regulations, management and treatment practices, as well as future perspectives for more sustainable nutrient-rich waste management on board of ships. As such, this chapter may point out opportunities to reverse human-accelerated eutrophication of marine ecosystems.

2. Sources of nutrient-rich wastes and wastewaters on board of ships

2.1 Food waste

The international maritime organization (IMO) defines ship-generated food waste as spoiled or unspoiled food substances containing fruits, vegetables, dairy products, meat products, and food scraps [14]. Large vessels (cargo and cruise ships) generally classify these residues into soft organic food waste (ex. peels and leftovers) and hard organic food waste (ex. bones), as well as packaging [15]. The quantity and composition of ship-generated food waste depends on a variety of parameters such as the ship type, the sorting strategy, the geographical area, the choice of the menu, etc. Cruise ships typically create the highest amount of food waste, with values up to 3.5 kg/person/day [8]. This waste type has an average dry weight (DW) and organic carbon content in the range of 22–38% and 46–60%, respectively, and an average nitrogen and phosphorus content in the range of

8.4–43 g/kg DW and 4.2–8 g/kg DW, respectively [16]. The management of food waste is a major concern on ships since the wet material is subject to fast degradation with odor pollution as a result. Hence, proper and efficient waste management strategies must be put in place.

2.2 Sewage

As for land-based wastewaters, ship-generated wastewaters are generally classified into black and gray water. Black water is sewage generated by toilets and medical facilities, while gray water is generated by showers, washing machines, and dish washers. Sewage on ships is generally more concentrated (about 2–3 times) than its land-based equivalent due to water-saving measures on board [12]. Studies Butt [17] and Svaetichin [18] estimated the volume of wastewater generated by cruise ships (2000–3000 passengers) in the range of 550–800 m^3/day of gray water and 110–115 m^3/day of black water. The release of nutrients into ship-generated sewage water is estimated at 12–15 g/person/day for nitrogen and 3–5 g/person/day for phosphorus [12, 19]. These nutrients that are responsible for eutrophication constitute a large proportion of the sewage water, hence to avoid harmful environmental impacts sewage should be properly stored and/or treated.

3. Ship-generated waste regulations

Ship-generated waste discharge is regulated by the MARPOL convention, i.e. the International Convention for the Prevention of Pollution from Ships adopted in 1973 by the IMO specialized agency of the United Nations and the global regulator of shipping [20]. According to Annex V of [21], food waste is an organic material categorized as garbage. An important feature of the Annex is the complete ban imposed on the disposal into the sea of all forms of plastics. For food waste itself, discharge following comminution or grounding of the residues is generally permitted at a distance of more than three nautical miles from the nearest land. However, in special designated areas such as the Baltic Sea, a distance of more than 12 nautical miles from the nearest land must be respected [22]. Discharge of not comminuted or grounded food waste is prohibited in special areas, but is allowed outside special areas at a distance of more than 12 nautical miles from the nearest land [22]. Ships are also allowed to shred and store their food waste on board for delivery at port reception facilities (PRFs) where it could potentially be collected and valorised. However, for ships that travel internationally, the food waste is classified as international waste and must therefore be eliminated. In the European Union for example, international food waste is considered as " high-risk category 1 animal by-products " [23, 24].

Sewage water discharge is regulated according to MARPOL Annex IV [25]. Ship-generated black water can generally, i.e. in non-special areas, be directly released into the open sea at a distance of 12 nautical miles from the nearest land. Sewage that is comminuted and disinfected using an approved system can be discharged into the sea at a distance of three nautical miles from the nearest land. General sewage water effluent standards prior to discharge into the sea are provided in **Table 1**. In special areas such as the Baltic Sea, additional sewage discharge restrictions apply. As such, for cruise ships operating in special areas, special limitations for nitrogen and phosphorus discharge have recently been established: maximum effluent concentrations of 20 mg/L (or 70% reduction) for nitrogen and 1 mg/L (or 80% reduction) for phosphorus. There are also voluntary initiatives in the shipping industry. As such, the European Cruise Council implemented the Agreement on

	Fecal coliforms (/100 mL)	Suspended solids (mg/L)	pH	Biochemical oxygen demand (BOD5, mg/L)	Chemical oxygen demand (COD, mg/L)
Treatment plant installed before 1/1/2010	250	50 if tested ashore (100 if tested on-board)	—	50	—
Treatment plant installed after 1/1/2010	100	35	6–8.5	25	125

Table 1.
Sewage water effluent standards prior to discharge into the sea at more than 3 nautical miles from the nearest land (MARPOL, Annex IV) [25].

	Food waste	Sewage water
General	Food waste can be comminuted/grounded and discharged into the sea at least 3 nautical miles from the nearest land if the ship is en route. Discharge of not comminuted or grounded food waste is allowed at a distance of more than 12 nautical miles from the nearest land if the ship is en route. Delivery of comminuted food waste to port reception facilities is allowed. Discharge of plastics is not allowed.	Sewage discharge into the sea is prohibited except if the ship has an approved sewage treatment plant or in the case that the ship discharges comminuted and disinfected sewage using an approved system at a distance of more than 3 nautical miles from the nearest land. Not comminuted or disinfected sewage needs to be discharged at a distance of more than 12 nautical miles from the nearest land when the ship is en route and proceeding at not less than 4 knots, and the rate of discharge should be approved.
Special area	Food waste can be comminuted/grounded and discharged at least 12 nautical miles from the nearest land. Delivery of comminuted food waste to port reception facilities is allowed.	From 2019 on, all new passenger ships must either treat nitrogen and phosphorus in black water or leave black water at port reception facilities for treatment in wastewater purification systems. Untreated black water cannot be pumped into the ocean. There are no special limitations for gray water.

Table 2.
Simplified overview of MARPOL regulations regarding food waste, black water and gray water for passenger ships [21, 22, 25].

Discharges in the Baltic Sea. This agreement declares that its members will stop releasing wastewater into the Baltic Sea and instead deliver it to port reception facilities without a special fee. Further, it must be remarked that wastewater treatment still generates a concentrated residual product, i.e. the sewage sludge, which also needs to be treated or delivered to port reception facilities [12].

A simplified summary of the MARPOL regulations applicable to organic waste sources is provided in **Table 2**.

4. Management and treatment practices

The management of ship-generated food waste is typically specific to ship policy. Direct discharge of food waste into the sea following grinding or comminution is for sure the cheapest and most straightforward method. However, since this practice is associated with environmental issues, discharge is not always possible according to regulations (**Table 2**). Moreover, some ship owners prefer to opt for more sustainable solutions on a voluntary basis. Alternative food waste management strategies currently applied include comminution, shredding or grinding of the waste source on board, followed by collection in bins and delivery to port reception facilities for disposal or further treatment. However, storage of food waste on board is challenging for multiple reasons: 1) It can carry diseases or pests and hence needs to be stored in covered containers, 2) It can involve large volumes with risk of putrefaction and odors, hence drying of food waste is recommended, as well as storage in a cooled room, 3) Larger ships should distinguish between soft and hard organic waste for separate storage and treatment; The hard organic waste and packaging is generally stored in bags or bins and delivered to port reception facilities. 4) International food waste needs to be handled differently from domestic food waste because of the risk of spreading diseases [8]. An overview of current common ship-generated food waste management practices is provided in **Figure 1**.

Ship-generated sewage needs to be treated onboard prior to discharge at more than 3 nautical miles in order to respect the regulations provided in **Table 1** or delivered to port reception facilities. In the latter case, the sewage is collected in a storage tank on board and chemicals are added for odor and color removal, as well as disinfection, prior to delivery to the reception facility. Sewage storage strategies vary depending on the type of ship; some vessels store both black and gray water in the same tank. However, storing onboard is associated with difficulties such as limited storage space, next to odor and pest control, so ships need to go to land at regular intervals if direct release of wastewater into the sea is not possible [8, 26].

Most large ships (cargo and cruise ships) have an approved treatment system onboard allowing them to discharge the wastewater into the sea following proper treatment. In a first stage, a pre-treatment such as screening to remove grit and debris is typically applied. Next, an aerobic biological treatment step (activated sludge) is generally applied to remove solids, biological oxygen demand (BOD5) and some nitrogen. Finally, a disinfection step is applied, typically using chlorine. A simplified schematic overview of the conventional treatment process is provided in **Figure 2**.

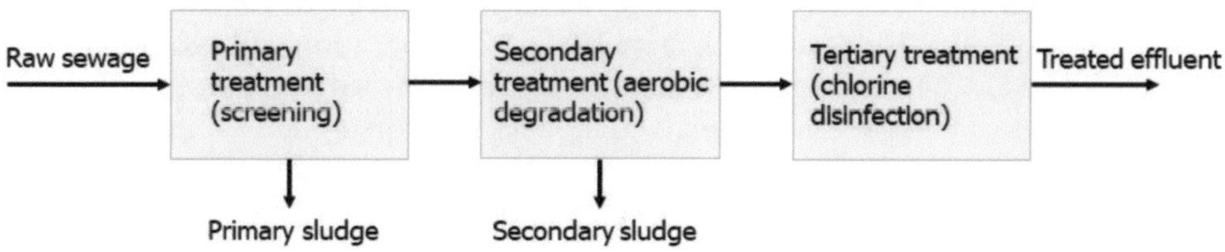

Figure 1.
Overview of common ship-generated food waste management practices. The dotted line indicates a common practice that is prohibited in special areas. NM: Nautical miles.

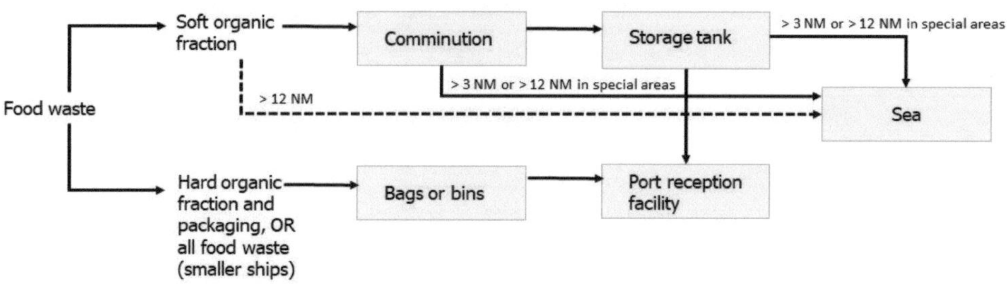

Figure 2.
Conventional on-board treatment process for ship-generated sewage on large ships adapted from 8, 15.

More advanced wastewater treatment systems are also applied involving improved screening, biological treatment (e.g. using membrane bioreactors), solids separation (e.g. using filtration and flotation) and disinfection (e.g. using ultraviolet light) [27]. The conventional wastewater treatment systems target the removal of suspended solids, BOD5 and pathogens, but typically only remove 58–74% of ammonia and 41–98% of phosphorus [26]. With the new regulations on nutrient discharges (nitrogen and phosphorus) in special areas, a variety of more advanced systems have been proposed to remove these nutrients down to the new discharge levels (**Table 2**). According to Helcom [11], there are currently 52 different wastewater purification systems on the market that meet these special area requirements.

5. Perspectives

While food waste and sewage sludge are often still discharged into the sea, these resources could potentially be valorised in a sustainable way. The study of [28] presents potential valorisation scenarios for domestic ship-generated organic wastes. Five different scenarios were proposed in this study including: 1) Composting on board of the ship, 2) Centralized composting, 3) Composting at the port, 4) Centralized anaerobic digestion, and 5) Anaerobic digestion at the port. Composting involves the aerobic degradation of organic waste in the presence of oxygen to produce an organic soil amendment (the compost). Anaerobic digestion involves the anaerobic degradation of organic waste to produce biogas (bioenergy) and biofertilizer (digestate). These ship-generated organic waste valorisation scenarios were compared in terms of their advantages and disadvantages, the required equipment, and associated costs and revenues [28]. The study concluded that the optimal scenario will depend on 1) the amount of organic waste produced by ships and available at the port, 2) the proximity of an existing centralized treatment plant, and 3) the potential market value and opportunities for composts and digestates produced in the area.

In order to facilitate case-specific decision-making, a decision-support software tool for optimal selection of organic waste management strategies is under development by the first author's research team in close collaboration with experts in geomatics and Quebec industry (www.optim-o.com). The software tool combines a multidimensional database, mathematical models, and a geographical information system to facilitate the development and selection of optimal scenarios. The scope includes the generation and collection of organic waste, the treatment of the waste

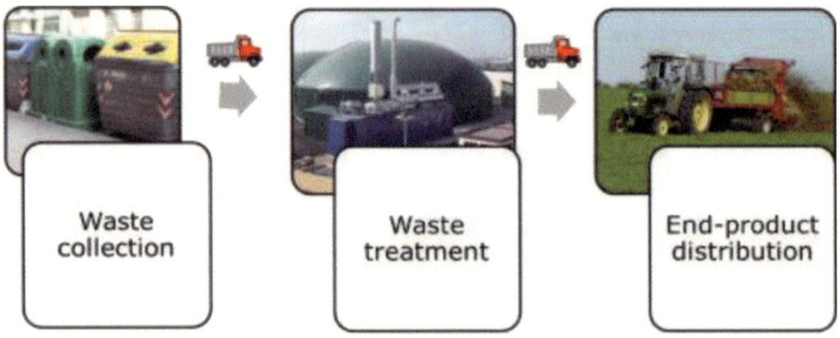

Figure 3.
Scope of the optim-O decision-support software tool for optimization of organic waste valorisation chains. Technical, spatiotemporal, environmental, economic, legal and social aspects of waste valorisation are taken into account. Images can be reused under the creative commons license agreement.

through biomethanation, composting and/or nutrient recovery, and the distribution of the end-products such as biogas, digestate, compost and recovered mineral fertilizers (**Figure 3**). As such, the entire valorisation chain can be optimized, taking into account environmental (e.g. greenhouse gas emissions), economic (e.g. operating costs), technical (e.g. process operational conditions), legal (e.g. fertilizer application restrictions), social (e.g. traffic nuisance) and spatiotemporal (e.g. transport distance and route) aspects. Although the tool was initially developed for land-generated organic waste management, future research will look at applications in the maritime sector.

Anaerobic digestion may offer a valuable solution for the valorisation of international organic waste, which is currently eliminated as required by regulation. International waste could be treated in a separate digestion unit at the port, or a small-scale system could be installed on the ship. In this way, the biogas produced from this international waste source could be valorized, while the residual digestate could be disposed of if valorisation would not be possible according to international waste regulations. This perspective will be further explored with the Canadian food inspection agency in the near future.

Finally, conventional wastewater treatment systems on board of ships target the removal of solids and pathogens, and some newer systems also target the removal of nitrogen and phosphorus. Future work will look at the recovery of these valuable nutrients as concentrated fertilizers products or other bioproducts instead of their removal [29]. Strategies such as nitrogen stripping-scrubbing to produce ammonium sulfate liquid fertilizer solution [29], the precipitation of struvite ($MgNH_4PO_4{:}6H_2O$) fertilizer [29] or the application of hybrid anion exchange nanotechnology for phosphorus recovery [30] could provide valuable solutions. An integrated process for nutrient recovery on board of ships will be aspect of research by the first author's research team. As such, sustainability in the maritime sector can be further be improved.

6. Conclusions

Ship-generated nutrient-rich wastes and wastewaters have detrimental impacts on marine ecosystems through eutrophication and ocean deoxygenation. Food waste is currently often discharged into the open sea without any treatment at more than 12 nautical miles from the nearest land. Sewage water can also be discharged into the sea without any treatment at more than 12 nautical miles. However, treatment (solids removal + disinfection) is required for discharge between 3 and 12 nautical miles from the nearest land. Large ships typically have an approved sewage treatment system on board. In special areas, untreated sewage water cannot be discharged into the ocean, and special discharge limits apply for nitrogen and phosphorus. Hence, in recent years, more advanced treatment systems for nutrient removal on board of ships have been proposed. Both food waste and sewage can also be stored on board and delivered to port reception facilities for disposal or treatment. Future research will look at the valorisation of these waste and wastewater sources through composting, anaerobic digestion and/or nutrient recovery in order to further improve sustainable resource management in the maritime sector.

Acknowledgements

The first author is funded by the Natural Science and Engineering Research Council of Canada through the award of an NSERC Discovery Grant

(RGPIN-2017-04838). Céline Vaneeckhaute holds the Canada Research Chair in Resource Recovery and Bioproducts Engineering.

Conflict of interest

The author declares no conflict of interest.

Author details

Céline Vaneeckhaute
Research Team on Green Process Engineering and Biorefineries (BioEngine), Chemical Engineering Department, Université Laval, Québec, Canada

*Address all correspondence to: celine.vaneeckhaute@gch.ulaval.ca

References

[1] EPA. National Pollutant Discharge Elimination System (NPDES). Permit Limits – Nutrient Permitting [Internet]. 2021. Available from: Permit Limits-Nutrient Permitting | National Pollutant Discharge Elimination System (NPDES) | US EPA [Accessed: 2021-05-02]

[2] JRC European Commission. Nutrient discharge from rivers to seas for year 2000. JRC Scientific and Technical Reports. 2009. Available from: JRC Publications Repository (europa.eu) [Accessed: 2021-05-02]

[3] Breitburg D, Levin LA, Oschlies A, Grégoire M, Chavez FP, Conley DJ, Garçon V, Giblert D, Gutiérrez D, Isensee K, Jacinto GS, Limburg KE, Montes I, Naqvi SWA, Pitcher GC, Rabalais NN, Roman MR, Rose KA, Seibel BA, Telszewski M, Yasuhara M, Zhang J. Declining oxygen in the global ocean and coastal waters. Science. 2018;359:6371. DOI: 10.1126/science.aam7240

[4] IUCN. Ocean deoxygenation. 2019. Available from: Ocean deoxygenation | IUCN [Accessed: 2021-05-02]

[5] Schmidtko S, Stramma L, Visbeck M. Decline in global oxygen content during the past five decades. Nature. 2017;542:335-339. DOI: 10.1038/nature21399pmid:28202958

[6] Steckbauer A, Duarte CM, Carstensen J, Vaquer-Sunyer R, Conley DJ. Ecosystems impacts of hypoixia: Thresholds of hypoxia and pathways to recovery. Environ. Res. Lett. 2011;6:025003. DOI: 10.1088/1748-9326/6/2/025003

[7] EPA. Nutrient Pollution. Harmful Algal Blooms. 2021. Available from: Harmful Algal Blooms | Nutrient Pollution | US EPA [Accessed: 2021-05-02]

[8] Vaneeckhaute C, Fazli A. Management of ship-generated food waste and sewage on the Baltic Sea: A review. Waste Manage. 2020;12-20. DOI: 10.1016/j.wasman.2019.10.030

[9] Jägerbrand AK, Brutemark A, Sveden JB, Gren IM. A review on the environmental impacts of shipping on aquatic and nearshore ecosystems. Sci. Total Environ. 2019;695:133637. DOI: 10.1016/j.scitotenv.2019.133637

[10] Brida JG, Zapata-Aguirre S. Cruise tourism: Economic, socio-cultural and environmental impacts. Int. J. Leis. Tour. Market. 2009;1:205-226. DOI: 10.1504/IJLTM.2010.029585

[11] Helcom. Ensuring safe shipping in the Baltic. Helsinki Commission, Finland. 2009. Available from: Ensuring safe shipping in the Baltic (helcom.fi) [Accessed: 2021-05-02]

[12] Huhta H-K, Rytkönen J, Jukka S. Estimated nutrient load from wastewaters originating from ships in the Baltic Sea area. VTT Technical Research Centre of Finland, Otakaari, Finland. 2007. Available from : Estimated nutrient load from waste waters originating from ships in the Baltic Sea area (vttresearch.com) [Accessed: 2021-05-02]

[13] IMO. Particularly Sensitive Sea Areas. 2021. Available from: https://www.imo.org/en/OurWork/Environment/Pages/PSSAs.aspx [Accessed: 2021-06-01]

[14] MEPC. Designation of the Baltic Sea area as a particularly sensitive sea area. Marine Environ. Protect. Commit. 2005;136(53). Available from: http://www.rise.odessa.ua/texts/MEPC136_53e.php3 [Accessed: 2021-05-02]

[15] CE Delft. The management of ship-generated waste on-board ships. CE Delft for European Maritime Safety

Agency, Delft. 2017. Available from: The Management of Ship-Generated Waste On-board Ships - CE Delft - EN [Accessed: 2021-05-02]

[16] Wilewska-Bien M. Management of ship-generated food waste: Illustrated from the Baltic Sea perspective [thesis]. Göteburg, Sweden: Chalmers University of Technology; 2017.

[17] Butt N. The impact of cruise ship generated waste on home ports and ports of call: A study of Southampton. Marine Policy. 2017;31(5):591-598. DOI: 10.1016/j.marpol.2007.03.002

[18] Svaetichin I. Cruise ship generated waste in the Baltic Sea - A study from the port's point of view on a possible updated waste management system [thesis]. Helsinki, Finland: University of Helsinki; 2016.

[19] USEPA. Cruise Ship Discharge Assessment Report. United States Environmental Protection Agency, Oceans and Coastal Protection Division, Washington, DC. 2008. Available from: Cruise Ship Discharges and Studies | Vessels, Marinas and Ports | US EPA [Accessed: 2021-05-02]

[20] IMO. Special areas under MARPOL, International Maritime Organization. 2019. Available from: List Of Special Areas Under MARPOL | Mariner's Circle (marinerscircle.com) [Accessed: 2021-05-02]

[21] MARPOL. Annex V: Regulations for the prevention of pollution by garbage from ships. 1988. Available from: Prevention of Pollution by Garbage from Ships (imo.org) [Accessed: 2021-05-02]

[22] MARPOL. Simplified overview of the discharge provisions of the revised MARPOL Annex V, which entered into force on 1 March 2018. 2018. Available from: https://wwwcdn.imo.org/localresources/en/OurWork/Environment/Documents/Simplified overview of the discharge provisions of the revised MARPOL Annex V.pdf [Accessed: 2021-05-02]

[23] European Commission. Directive 2000/59/EC of the European Parliament and of the Council of 27 November 2000 on port reception facilities for ship- generated waste and cargo residues. 2000. Available from: EUR-Lex - 32000L0059 - EN - EUR-Lex (europa.eu) [Accessed: 2021-05-02]

[24] European Commission. Regulation (EC) No. 1069/2009 of the European Parliament and of the Council laying down health rules as regards animal byproducts and derived products not intended for human consumption and repealing Regulation (EC) No. 1774/2002 (Animal by-products Regulation). 2009. Available from: EUR-Lex - 32009R1069 - EN - EUR-Lex (europa.eu) [Accessed: 2021-05-02]

[25] MARPOL, 2003. Annex IV: Regulations for the prevention of pollution by sewage from ships. Available from: Prevention of Pollution by Sewage from Ships (imo.org) [Accessed: 2021-05-02]

[26] Granhag L, Wilewska-Bien M, Andersson K. Rapporten Kartläggning av näringstillförsel från sjöfart till Östersjön är finansierad av Stiftelserna Thurséus Forskarhem och BalticSea2020 [report]. Göteborg, Sweden: Chalmers tekniska högskola; 2018.

[27] Koboevic Z, Kurtela Z. Comparison of marine sewage treatment systems. https://bib.irb.hr/datoteka/570916.COMPARISON_OF_MARINE_SEWAGE_TREATMENT_SYSTEMS.pdf

[28] Vaneeckhaute C, Darveau O. Current state and potential valorisation of ship-generated organic waste in Quebec, Canada. Waste Manage. 2020;118:62-67. DOI: 10.1016/j.wasman.2020.08.009

[29] Vaneeckhaute C, Lebuf V, Michels E, Belia E, Tack FMG, Vanrolleghem PA, Meers E. Nutrient recovery from bio-digestion waste: Systematic technology review and product classification. Waste Biomass Valor. 2017;8(1):21-40. DOI: 10.1007/s12649-016-9642-x

[30] Ownby M, Desrosiers DA, Vaneeckhaute C. Phosphorus removal and recovery from wastewater via hybrid ion exchange nanotechnology: A study on sustainable regeneration chemistries. Npj Clean Water. 2021;4:6. DOI: 10.1038/s41545-021-00104-7

Chapter 5

Perspective Chapter: Smart Maintenance in Modern Ship Engineering, Design and Operations

Saulo Vasconcelos and Paulo Vasconcelos

Abstract

Marine Engineers are forged to face the most complex adversities of the maritime environment, whether in cabotage, maritime support, fluvial or on oil and gas rigs. Usually, the Marine Engineer is responsible for the production of potable water, lubricating oil and fuel system, sanitary and cooling system, propulsion system, electricity generation and others. This chapter will use quality tools, including the dart of Vasconcelos to quickly find the root cause of the problems. Troubleshooting is part of day-to-day job of an experienced Engineer, and he knows what kinds of techniques to use to troubleshoot and solve problems in an organized, quick and easy way.

Keywords: Marine Engineer, smart maintenance, troubleshooting, quality tools, dart of Vasconcelos

1. Introduction

The recent trend to design smart modern ships, more efficient with a high level of automation and processes control, this introduction of new technologies and changes in manning, requires from the Marine Engineers expertise to quickly solve the problems presented in the routine on board of modern merchant ships and oil rigs.

Marine Engineers are forged to face the most complex adversities of the maritime environment, whether in cabotage, maritime support, fluvial or on oil and gas rigs. Usually, the Marine Engineer is responsible for the production of potable water, lubricating oil and fuel system, sanitary and cooling system, propulsion system, electricity generation and others.

It is especially important to know the operating parameters of the systems and equipment, in order to give the correct diagnosis to his equipment and to know exactly which “medication” needs to be applied to solve the problem.

This chapter will use quality tools, the dart of Vasconcelos illustrated in **Figure 1** [1] to quickly find the root cause of the problems. Troubleshooting is part of day-to-day job of an experienced Engineer.

IntechOpen

Figure 1.
Dart of Vasconcelos.

2. What is troubleshooting?

Troubleshooting can be defined as a logical search based on the symptomatic of a system or equipment, about what could be causing a particular problem, what would be its origin and how we could solve it to make the system or equipment operational again. One of the most important things to do an efficient troubleshooting is have a considerable knowledge from the subject in question, that can be obtained from the job experience and/or from a database (often from the system or equipment catalog itself) which provides a basis for the failure analysis that occurred.

Therefore, in summary, troubleshooting can be viewed as a careful way, which allows to accurately identify all the details of the system and, from that it is possible to make small changes in order to arrive at the root cause of the failure, as well as the to analyze each of the listed hypotheses to see if they can solve the problem effectively.

2.1 Dart of Vasconcelos

The basics steps of a good troubleshooting consist in identify the faced problem, establish a theory for the problem, test the theory, establish an action plan, and implement the solution. It is also possible to divide the troubleshooting sequence in define the problem through the ***Symptomatic***, do a ***Logical Search***, creating hypotheses, do the ***Contestation***, reducing the hypotheses and, finally, proceed with the ***Solution*** to solve the problem, where the hypotheses are tested and confirmed. Therefore, we can simplify that sequence through the of "Dart of Vasconcelos" illustrated in **Figure 1**.

Symptomatic:

Definition of the problem. It is essential to understand the system affected and what should be analyzed. It consists on the title of the troubleshooting.

Example:

1. Main engine #02 running with low fuel oil pressure. Title could be: "**Low fuel pressure oil on main engine #02**".

Logical Search:

Create hypotheses. Consist in analyze what systems or equipment are being affected by the problem and in to do the temporal analysis of the problem, the timeline, to be able to formulate hypotheses for the problem. In other words, create hypotheses to know exactly what should be inspected in each situation. Obviously, this will require good knowledge of the equipment and system in question, as well as the help of drawings, catalogs and manuals to understand the operation of what is being analyzed.

Example:

1. *System affected:* Fuel Injection System.

2. *Temporal analysis:* The problem occurred after maintenance on the fuel feed pump.

Creating Hypotheses:

1. Dirty or blocked fuel filters on engine.

2. Dirty or blocked strainer for fuel feed pump.

3. Fuel injector problems.

4. Bad fuel conditions.

5. Fuel feed pump wrong internal gear installed.

6. Fuel feed pump with clearance in internal gear.

Contestation:

Reduce hypotheses. At that phase, you should contest the hypotheses created with the intention of reduce at the maximum the possibility of causes for the problem. Discuss hypotheses with all involved on the troubleshooting task, making use of manuals, schematic drawings, maintenance history and other relevant information, until minimize the hypotheses.

To solve the problem, is essential eliminate hypotheses.

Reducing Hypotheses:

1. Dirty or blocked fuel filters on engine. Not possible, once engine was running with a good fuel pressure before to be stopped for pump maintenance.

2. Dirty or blocked strainer for fuel feed pump. Not possible due fuel pressure on suction and discharge lines of the pumps was good before pump maintenance.

3. Fuel injector problems. Not possible, Injectors were replaced recently, and no parameter indicated a injector problem when engine was running.

4. Bad fuel conditions. Not possible once other engine was running at that time and using the same fuel.

5. Fuel feed pump wrong internal gear installed. Possible, once can compromise the fuel flow and/or pressure.

6. Fuel feed pump with clearance in internal gear. Possible, once can compromise fuel flow and/or pressure.

Solution:

Solve the problem.

On that last step, once minimize the hypotheses, it is time to test each one, always starting from the easier one, to confirm and validate the hypotheses and finally find the solution for the problem. The correct diagnosis is extremely important. It is a critical part of the troubleshooting, once a wrong action can make you develop other undesirable problems and do the task harder, spending more time and producing emotional stress among the team and compromising the chance to quickly find the real problem. Remember to be organized and caution during the hands on for the tests.

Therefore, discovered the problem, the corrective action will be taken to put back on service the system/equipment. Remember to always log the problem with a report, thereby increasing the database to fix any such problems that may occur, and thus helping the team solve the problem in the future when you are not around.

Indeed, as the dart game has the intention to hit the bull's-eye, the Dart of Vasconcelos has the intention to hit it and solve the problem, which is the main target.

2.2 Practical example of troubleshooting

A particular ship began to show a significant loss of lubricating oil in one of its main electric power generation engines, after maintenance of replacing the liners and sealing rings of the corresponding pistons. In that case, the troubleshooting would lead us, assisted of the Dart of Vasconcelos, to reason as follows: (Definition of the problem: Symptomatic) what could be causing this significant loss of lubricating oil (Create hypotheses: Logical Search)? How to determine the source of the problem (Reduce hypotheses: Contestation) and how to act to make the system/equipment operational again (Solve the problem: Solution)? [1].

Symptomatic:

Significant loss of lubricating oil. Title: "**Loss of lubricating oil on engine**".

Logical search:

1. Lube oil purifier discharging excessive lubricating oil to sludge tank.

2. Lube oil leaking from pumps or hoses.

3. Lubricating oil passing through combustion chamber due seal ring damaged.

Contestation:

Reducing Hypotheses:

1. Lube oil purifier discharging excessive lubricating oil to sludge tank. Not possible, once was monitored the sludge tank level and tank level was according with expected.

2. Lube oil leaking from pumps or hoses. Not possible. Was inspected all engine for external leaking and nothing unusual was found.

3. Lubricating oil passing through combustion chamber due seal ring damaged. Possible, once the seal rings were replaced during maintenance. It could be allowing part of the lubricating oil to pass into the combustion chamber and be burned together with the engine fuel.

Solution:

Solve the problem:

To validate the possible hypothesis, is necessary to test to confirm. One way to identify with lube oil has been burned together fuel is to observe the color of the exhaust gases from the engine. If it is bluish gray in color, this means that the lubricating oil is probably being burned together with the fuel.

Another easy test to do is a compression test performed on the cylinders that passed for the maintenance. The cylinder with low compression pressure should certainly be the cylinder with the problem.

Therefore, once done the correct diagnosis, proceed with the corrective action. Carry out the inspection on the cylinder with low compression pressure and confirm the damaged seal ring and then, replace it. Register the problem making a report, once it increases your database of troubleshooting for any similar problems that may happen and assist the team to solve problems when you are not around.

2.3 Ishikawa diagram/fishbone diagram

The Ishikawa diagram, or Fishbone diagram, as it is also known, is a cause-and-effect diagram that helps to find the root causes of a problem, analyzing the factors that involve the execution of the process. This method was created by the Japanese Kaoru Ishikawa in the 60's and it's highly used today in the most diverse industries. The main factor of the diagram is to consider all aspects that may have generated the problem, thus making it easier to organize ideas to arrive at a solution for the problem.

It is a fact that every problem has specific causes. Analyzing each cause and verifying the possibility that it is the root of the problem is the key to finding a solution. It is necessary, whenever possible, to analyze the simplest and most probable causes of the problem, in order to save time and avoid hard and exhausting work during troubleshooting.

The great relationship of the Ishikawa Diagram with the spine of a fish comes from the fact that we can consider the spines as the causes of the observed problems, which will contribute to the discovery of their effect. The Ishikawa Diagram has great applicability in different contexts and in different ways, among them, the use stands out:

1. To view the main and secondary causes of a problem (effect).
2. To broaden the vision of the possible causes of a problem, seeing it in a more systemic and comprehensive way.
3. To identify solutions, raising the resources available by the company.
4. To generate process improvements.

2.4 How to make an Ishikawa diagram?

In order to carry out the Fishbone diagram, we must take the following steps:

1. Define the problem causes (effect) to be analyzed.
2. Draw a horizontal arrow pointing to the right and write the problem inside a rectangle located at the tip of the arrow.
3. Brainstorm to raise possible causes that may be causing the problem. To do this, try to answer the following question: "Why is this happening".

4. Divide the causes identified into categories, for example: machine, manpower, method, mother-nature, measure and materials or in a way that is most consistent with the problem analyzed and the context of your company.

5. Right after, define the sub-causes, that is, the factors that caused that cause to happen.

The main categories represented in each of the spines are also known as the "6 M", referring to the following items:

1. Manpower (People): involved in the process.

2. Method: The way the process is performed. What are the procedures used, rules and requirements.

3. Machine: The equipment involved in the process.

4. Measurement: Analyze the metrics used. Analyze all the data generated from the process looking for failures.

5. Material: Evaluate all material used in the process, parts, for example.

6. Mother-Nature (Environment): Analyze the conditions of the place where the process is developed.

2.5 Benefits of using the Ishikawa diagram

It is easy to understand that the use of the diagram facilitates the visualization of the causes of problems and the effects caused by them. In addition, we cite as other benefits:

1. Facilitate team brainstorming, in order to facilitate the organization of ideas.

2. It helps to keep the focus of the whole team.

3. Show all possible causes at once, in an organized and segmented way.

4. It stimulates the solution of problems.

2.6 Example how to use Ishikawa diagram

As an example of how to use the Ishikawa diagram, let us see how it would be applied to identify high coolant consumption in an Engine Room in engines plant. As explained before, first you need to do the sketch of the fishbone diagram, writing the "6Ms" (cause) to find out the problem (effect). Once done that, each one of the "6Ms" should be analyzed in separate, showing what could be caused the high coolant consumption in the engines plant. It assists the brainstorm and make easier to find out the real cause of the problem. **Figure 2** shows exactly the diagram for the explained situation.

Therefore, as seen, the diagram facilities the visualization of what could be causing the problem. The next step would be to analyze among all the election causes, what would be the correct one to the problem, for after that, check how to solve it.

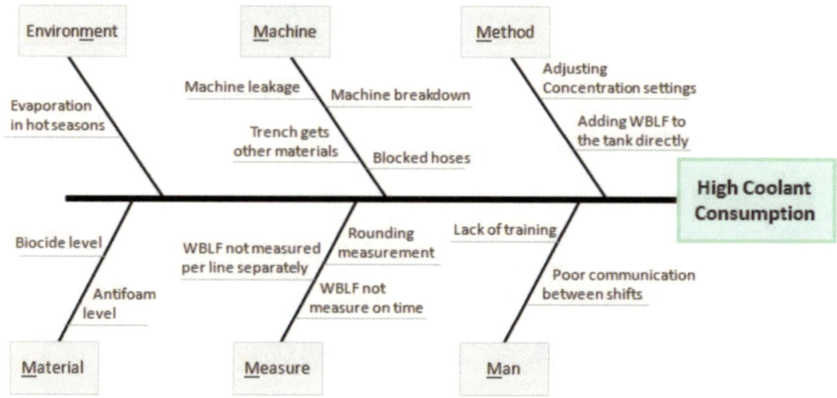

Figure 2.
Example of Ishikawa diagram. Causes that increased the coolant consumption in a manufacturing plant [1].

2.7 Difference between dart of Vasconcelos and Ishikawa diagram

Before all, it's particularly important to say that both methods are good and can be used for any problem of cause and effect. Ishikawa diagram use a wider view of the situation, while Dart of Vasconcelos is more objective, making faster to figure out what is causing the problem. Ishikawa is a type of diagram that anyone can participate, independent of being a maintenance team member, for example, a painter can say his opinion about one problem with his vision, making sense or not. The Dart, on the other hand, suggests that only people with affinity in the task raise opinions, so it becomes more objective and practical. Other difference is that Ishikawa shows the causes that could be causing the problem; the dart of Vasconcelos shows the causes and a way to solve them. Somehow, the person performing the troubleshooting can analyze which one method should be better applied for the present problem. To a better understanding, let analyze one situation using both methods.

2.8 Proposed problem

A specific diesel engine has been presented difficulty to start recently. Douglas, 2nd Marine Engineer, reported that this engine is with maintenance late and that has been presented an unusual and thick smoke from the exhaust gases. Therefore, could you find out the possible problem of the engine?

To proceed with that troubleshooting, let start using Ishikawa diagram and after the Dart of Vasconcelos. Therefore, building the fishbone for the problem, we have the following schematic illustrated in **Figure 3**.

Observe that the diagram has several causes for the problem (effect) and each one should be analyzed to confirm if it is a possible cause or not. At same time that diagram facilities the visualization of the causes, depending on the problem it can have many causes and sub-causes, becoming hard and lingering the task, once it seeks to look at the problem as a whole, and not objectively, based on the affected system and the on temporal analysis. Therefore, let us analyzing each cause:

1. Polluted or rarefied ambient air. Not possible. Ambient inspected and nothing unusual observed.

2. Incorrect parts used for maintenance. Not possible. Double checked part number to confirm parts used are correct.

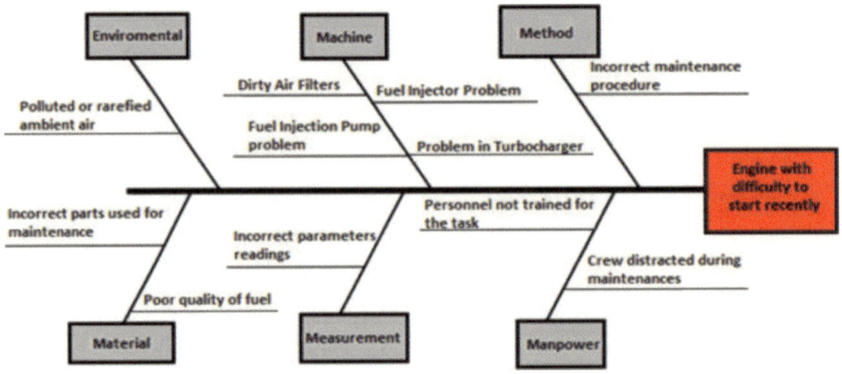

Figure 3.
Ishikawa's diagram for engine with difficulty to start recently [1].

3. Poor quality of fuel. Not possible. Collected a sample and saw that was in good conditions.

4. Air filters dirty. Possible.

5. Fuel injection pump problem. Possible.

6. Fuel injector problem. Possible.

7. Problem in Turbocharger. Possible.

8. Incorrect parameters readings. Not possible. Confirmed that instrumentation was working properly.

9. Personnel not trained for the task. Not possible. Experienced crew on the task.

10. Incorrect maintenance procedure. Not possible, Procedure used direct from the manufacture manual.

11. Crew distracted during maintenances. Possible.

Ishikawa diagram gives 5 possibilities for the problem. Analyzing each one we have:

1. **Dirty air filters**. Possible. An unusual and thick smoke from the exhaust gases reported by Engineer is a hint, once it indicates an uncomplete burn into the combustion chambers. Once maintenance is later, should be good check if air filters are not dirty and causing an obstruction of air.

2. **Fuel injection pump problem**. Possible. It can reduce the fuel pressure that pass to the fuel injectors.

3. **Fuel injector problem**. Possible. Same as dirty air filter, thick smoke from the exhaust gases reported by Engineer is a hint, once it indicates an uncompleted burn into the combustion chambers. Once maintenance is later, should be good check injectors for dirty nozzles or damaged. It is easier several injectors get dirty and occur the problem reported them injection pumps.

4. **Problem in Turbocharger**. Possible. If the temperature or pressure of the air into the turbocharger outlet is not according to the operational parameter, it indicates a bad condition of turbocharger. Other occurrence would be clogged turbocharger air filter element or any unsuitable cause in the air supply.

5. **Crew distracted during maintenances**. Possible. Besides crew team follows the manufacture manual procedures, if during the maintenance the team was distracted, they can forget some steps or not adjust as required injection timing, for example, that would cause same problem of dirty air filters, an uncompleted burn into the combustion chambers.

Therefore, electing the priority to be inspected (based on possibility and facility) we have inspect air filters, double check for the steps done with the crew during maintenance, inspect fuel injectors, inspect turbochargers, and inspect fuel injection pumps.

Now let us analyze the same proposed problem above using Dart of Vasconcelos:

Symptomatic: "specific diesel engine has been presented difficulty to start recently".

Definition of the problem: Engine with difficulty to start.

Logical Search:

1. System affected: Fuel system and air system.

2. Temporal analysis: Engine is with maintenance late.

Creating Hypotheses:

Based on symptomatic and the logical search, is possible significantly reduce the quantity of hypotheses, for example, poor quality of fuel will not be considered as a cause, once only one "specific" engine presented the problem, if was a common problem, certainly fuel would be a great possibility. The same happen to polluted or rarefied ambient air, using temporal analysis, nothing make reference about changes on the work engine area to consider it as a cause. Note that using this method, is possible to be more specific and surgical on the troubleshooting, saving time with possibilities that really are important to find out the problem.

Indeed, the following hypotheses for the question are:

1. **Dirty air filters**. Possible. An unusual and thick smoke from the exhaust gases reported by Engineer is a hint, once it indicates an uncomplete burn into the combustion chambers. Once maintenance is later, should be good check if air filters are not dirty and causing an obstruction of air.

2. **Fuel injection pump problem**. Possible. It can reduce the fuel pressure that pass to the fuel injectors.

3. **Fuel injector problem**. Possible. Same as air filter dirty; thick smoke from the exhaust gases reported by Engineers is a hint, once it indicates an uncompleted burn into the combustion chambers. Once maintenance is later, should be good to check injectors for dirty nozzles or damaged. It is easier several injectors get dirty and the problem reported in injection pumps occurs.

4. **Problem in Turbocharger**. Possible. If the temperature or pressure of the air into the turbocharger outlet is not according to the operational parameter, it

indicates a bad condition of turbocharger. Other occurrence would be clogged turbocharger air filter element or any unsuitable cause in the air supply.

Contestation:
Reducing hypotheses.

1. **Fuel injection pump problem**. Unlikely. would be necessary more of them one injection pump problem to the fault reported. Not so common several fuel injections pumps with problem at the same time.

2. **Problem in Turbocharger**. Unlikely. Other several problems would occur if the turbocharger problem happened, and certainly a characteristic and uncomfortable noise would be noted. The Marine Engineer also not reported any high temperature alarm in the turbocharges or cylinders, that would be common for high temperature in the outlet of the turbocharger.

Solution:
Solve the problem.
Therefore, the possibilities of the problems were reduced in two, dirty air filters or injector problems. Starting from the easier to inspect and solve, inspect air filters for clogged and replace it. For fuel injector, inspect all injectors for carbonization on nozzle, test the pressure of opening and how the spray, and check for inappropriate injection timing is.

In summary, using both methods will be possible find the causes to finally solve the problem. Vasconcelos method shows more objectivity, and Ishikawa looks the problem as a whole. The person in charge doing the troubleshooting will decide which method will be better for each situation.

2.9 Availability and reliability of equipment and plant operation

MTBF (mean time between failures) and MTTR (mean time to repair) are two indicators related to the availability of an Industrial Process.

2.9.1 MTBF

These are the periods of time that are lost while the machine is running and can be averaged using a formula. We have to apply the total performance time during a predetermined cycle under the number of failure that occurred during such time. According the Formula: MTBF = (Total Available Time – Waiting Time) / (numbers of shutdowns in the period of operation).

Example a freshwater generator designed to operate for 24 hours per day. Suppose the freshwater generator shutdown three times in the span of 30 days. The first shutdown occurred 48 hours from the start time and took 6 hours to repair. The second shutdown occurred 240 hours from the start time and took 4 hours to repair and the last shutdown 480 hours from start time and took 2 hour to repair before the freshwater generator was operating normally.

MTBF = (30daysx24hours - (6 hours + 4 hours+2 hour))/(3 shutdowns in 30 days of operations time) = 714 hours/3 = 236 hours or a mean time of 9 days and 20 hours between failures in the span of 30 days of the freshwater generator operation time.

With this conclusion, strategies can be created to face a problem gradually associated with the equipment.

2.9.2 MTTR

MTTR is calculated by applying the average time it takes to perform a repair after the failure episode. See the formula: MTTR = (Total repair time) / (number of failures). If you use the example above, you should get the following solution:

MTTR = (6 + 4 + 2) / 3 = 4 hours.

This solution establishes the average time that the equipment was stopped. Generating a relationship with the two indexes is the availability of this process.

AVAILABILITY = MTBF/(MTBF+ MTTR).

Therefore the availability of the freshwater generator in this process on board of a vessel is:

AVAILABILITY = 236 hours/(236 hours + 4 hours) = (236/240)x100 = 98,34%.

Thus, the lower the MTTR and the higher the MTBF, the more efficient the maintenance team will work.

Reliability of an equipment is related with the MTTF (mean time to failure), or failure rate, which expresses the probability of the equipment failures during a given period of time. It's normally applied to unrepairable devices, such as electronic devices and some relays with expected life of 300,000 cycles of operation or more or less 10 years of span life.

3. Troubleshooting based in real cases

In this part, is showed the questions based in real cases. Some of questions, before to be presented, have an initial introduction about the subject debated into the question, in order to help the reader to solve the problem. The reader may use one of the cause and effect methods spoken in the beginning of this chapter and use the ***hints*** of the questions for assistance him. Feel free to decide which one is better for you. Therefore, with no more conversation, let us initiate the troubleshooting!

3.1 Generator: Excitation

A specific, Dynamic Position vessel class #02, with electric diesel engines, configured with two medium voltage busbars, a bus tie breaker, four brushless diesel generators, two fixed pitch propellers and four tunnel thrusters illustrated in **Figure 4** [1].

Laura, an experienced Marine Engineer was ordered to prepare the engines, once the operations have finished at that port, so the vessel had to leave. The Diesel Generator 01 was on the busbar, with the bus tie breaker closed, as shown in **Figure 5**.

Therefore, Laura activated the Power Management System (PMS), so that the others diesel generators feed the busbars, thus making four generators on the busbars, to then drive the propulsion's loads. However, the engineer was surprised,

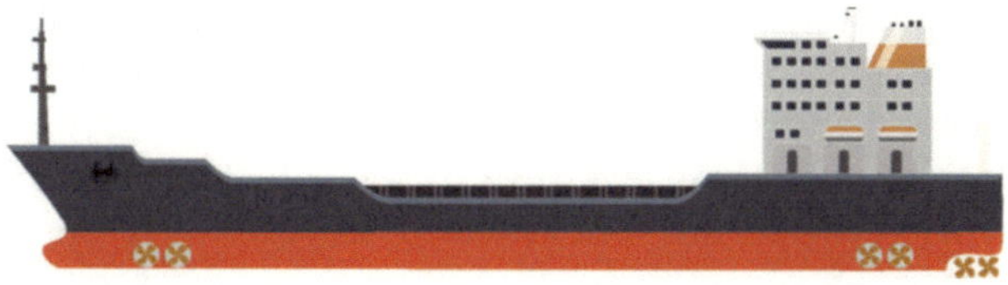

Figure 4.
Electric diesel vessel with 2 fixed thrusters and 4 tunnel thrusters [1].

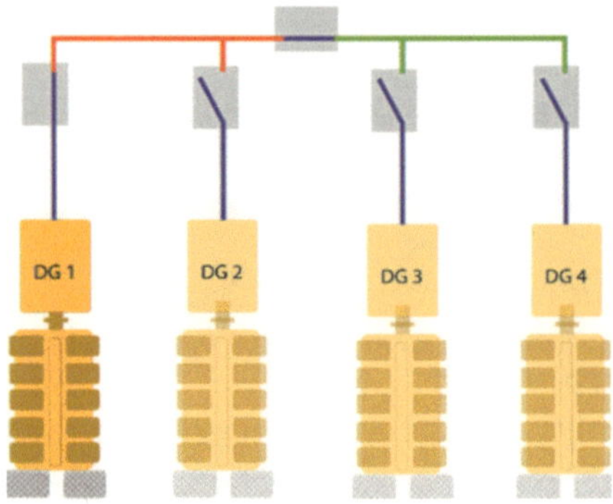

Figure 5.
Screen from power management system (PMS) during vessel operation at the port [1].

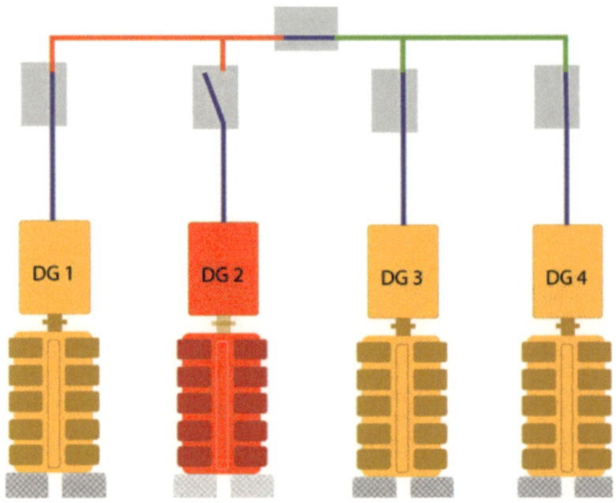

Figure 6.
Condition of screen from power management system (PMS) during the failure alarm [1].

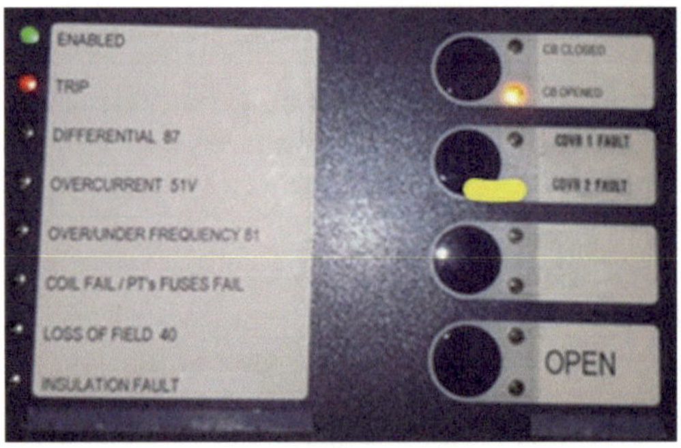

Figure 7.
Protection relay SEL – 7000G [1].

as generator #02 did not assume the load, generating several alarms, among them Automatic Voltage Regulator (AVR) fault, in the protection relay of generator #02, illustrated in **Figures 6** and 7.

Observing on the Bridge, **Figure 8**, the Captain called to Engine Room to understand why the generator had not assumed the load, once the company's procedure require that all generators should be on the busbars for the vessel to leave the port until navigation. Laura replied that she was investigating, doing her troubleshooting while, the Chief Engineer and the electrician were on their way to

Figure 8.
AVR switch button [1].

inspect the problem. Marine Engineer knew that this DP class #02 vessel had a lot of redundancy until arrives at the generator #02 busbars control cubicle, she recalled that each diesel generator had two AVRs, and then she quickly deselected the AVR #01 and selected the AVR #02, resetting the SEL 7000 bus protection relay.

At this moment, the Chief Engineer arrived at the Machinery Control Center (MCC) and Laura reported what she had done, promptly the Chief praised her and signaled her to continue the procedure that was underway with the electrician. Together, after all the conferences and resets, Laura, and the electrician Jair, issued the order for the PMS to start generator #02, which excited, synchronized, closed the circuit breaker on the busbars, and divided the load normally. Chief Engineer Nikola Tesla, asked to electrician Jair to look for a new AVR in the electric Warehouse, to replace the AVR #01, even so, Nikola thought about doing some tests.

Therefore, he requested that the Captain activate the tunnel thrusters above the port in two ways:

1. Slowly activating the thrusters with only generator #02 on the busbars.

2. Quickly triggering thrusters with generators #01 and # 02 on the portside busbars, illustrated in **Figure 9**.

Vessel responded well, there were no alarms, or the opening of the generator #02 circuit breaker on the busbars.

Thus, this vessel definitively left the port for navigation towards its offshore operations and the Chief returned to rest as well as the electrician, after troubleshooting, to then follow the next operations.

Gabriel, the First Engineer, assume his duty service, and after his normal round, started a revision service for a fuel oil purifier that Laura had started on her shift,

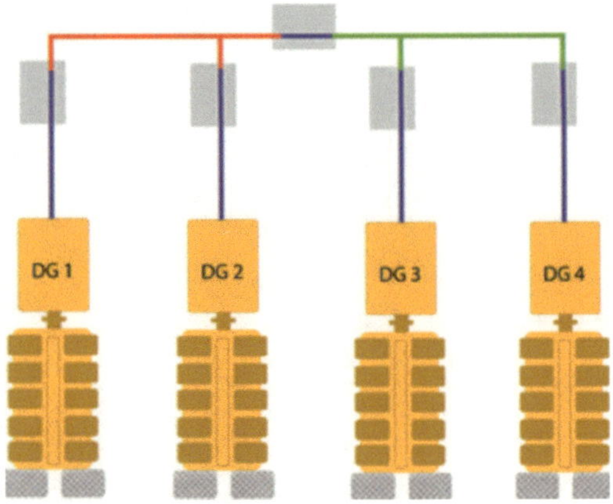

Figure 9.
Condition of main busbars during navigation to offshore operations [1].

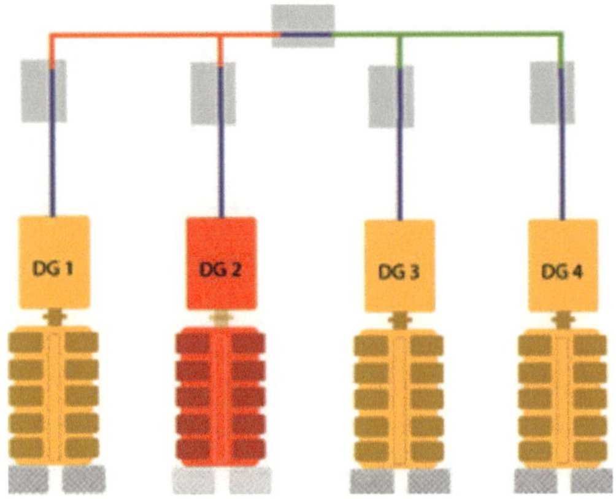

Figure 10.
Condition of main busbars during navigation after generator failure [1].

once the ship was navigating, and there would be enough time to finish that maintenance service.

As soon as the maintenance started, Gabriel observed a change in the charge regime of generator #02 due to the sound emitted by it, and soon there was an alarm. The engineer went to the MCC, noting the condition of generator on PMS screen, seeing that the Generator had disconnected from the busbars with a series of alarms, among them: "AVR Fault", again as illustrated in **Figure 10**.

At this moment, the electrician Jair said: "we need to see this, can be a serious problem!", because he knew that Laura had selected another AVR during troubleshooting, on the night before, and it would be very unlikely that both AVRs had failed.

Time was running out before the vessel reached the platform, so Gabriel decided to communicate Nikola, because only with three generators, there would be a reduction in contract speed, compromising the contract speed with the client. Since the vessel was equipped with PMS there was no blackout, the thrusters were automatically reduced and until then the Deck officers had not noticed any difference in the speed of navigation. So, when Chief Nikola arrives at MCC, Jair and Gabriel were already exchanging the AVR that Jair had already brought from the electric Warehouse. Nikola, very dissatisfied just observed what Gabriel and Jair were doing. Nikola is an experienced Chief Engineer and knew that the chance of two

AVRs failing together would be very low. Therefore, Nikola looked for the excitation / load control diagram of generator #02, while Jair and Gabriel were replacing the AVR. Nikola knew the order of events in which a generator needs to synchronize in parallel with another generator:

1. Order to depart the generator that must happen in a certain time.

2. Exciting (or increasing the AVR tension value), that should happen in a certain time.

3. When achieving the tension value of the busbars, which cause small rotation slips until the generators and the busbars have the same phase angle that must to happen in a certain time.

4. When achieving the same phase angle displayed in the synchronoscope which must be an order to close the circuit breaker.

5. There are number of attempts to close the circuit breaker. If the circuit breaker does not close, there is an engine stop order.

When the breaker closes, the loading generator increases the AVR current injection as well as fuel injection in the diesel machine while the generators that are giving load do the reverse process.

Telemetry Mesh called "loading sharing" communicates data to ensure that the Kilo Watts (KW) load values in both generators are equal, maintaining the AVR control process / fuel injection according to the electric load charges of the busbars.

Before Chief Nikola continued his analysis, Jair and Gabriel notified the order of the change in the AVR. Not ensured about the change done, they started the test. Generator #02 curiously connected on the busbars normally. Jair and Gabriel, were incredibly happy, shook hands and praised each other. Nikola thanked, but still dissatisfied said: Let us test!

Thus, at this moment, Chief Mate Fernanda, who had already been notified, started to take charge in several ways, thrusters against each other, all forward and all reverse and still generator #02 remained on the busbars. Jair and Gabriel were certain that everything was right, after all, in these tests, generator #02 remained on the busbars, the tests were severe, if anything that had to happen would happen during the tests. When Jair went to boast to the Chief, Nikola, who was focused on reading the electrical diagram, said to him: I am going up, call me when this generator leaves the busbars again! Skeptical and upset Gabriel said: Chief, it is settled, we have done the tests! Nikola replied with a friendly smile: Dear friend, thank you for your effort, your work was excellent, but call me again when this generator is disconnected from the busbars. Nikola turned his back and left MCC.

Gabriel told to Jair that the Chief was very suspicious fellow. Jair said that Nikola thinks him is the master of all the machines, he thinks that he invented the generator! After all, the electrician was Jair. Gabriel already calmer, said: Jair, you know how it is, right? Chief is Chief.

Thus, Jair said goodbye to Gabriel and went to change some lamps of the vessel, and Gabriel went to continue the service in the Purifier of diesel oil.

Time passed, the purifier was set up, Gabriel called Danilo, Oiler, to clean the environment of the purifier, when again generator #02, presented unusual noise and failed!

Surprised and incredulous, Gabriel said: what a Chief! Going up to the MCC where the generator #02 cubicle was located, he found the Chief with a Laser

Temperature Gun (LTG) in his hand, measuring temperature inside the generator #02 cubicle. Nikola called Jair and said, change this component, giving Jair an electric contactor, and pointing out on the board which one he wanted to be changed. Jair, feeling pressured by the constant defects, changed the component requested by Chief Nikola. Even when driving in different ways, thrusters against each other, all the power in front and all the power in reverse, generator #02 remained on the busbars. Nikola said: Now the problem is solved!

Therefore, could explain what was the Chief Nikola Tesla thinking and how he found out that the electric contactor was the real problem?

Hints: Field and excitation coils.

Solution:

To understand and solve this problem, is important observe the timeline analysis of the problem. Three AVRs failed. First AVR failure with Laura, it was replaced by it hot stand by AVR # 02 in 1 hour by Laura, and then failed again after 6 hours and was replaced by a new one in 1.5 hour of service, lastly, other AVR failed after 4 hours with Jair and Gabriel and again was replaced for a new one in 1.5 hour of service. The same failure happens with all AVRs, what easily suggest that the problem may not be this component and yes, another item common to for the AVRs. The purpose of an AVR, illustrated in **Figure 11** is to control the voltage of a generator by injecting current into the exciting coil. When you turn on a heater in your cabin, for example, the current in the generator armature increases and the magnetic field crossing the armature decreases due Lenz law, so the output voltage drops. To prevent this to occur, the AVR increases the current injection in the main exciter, increasing the magnetic field of the coil, consequently the field crossing the armature, and finally the output voltage to it set point.

When someone turns on a lamp, the AVR injects more current into the main exciter, when someone turns off, the AVR decreases the excitation, so the set point remains constant. If this control were manual, you would be in front of the electrical energy board all day increasing or decreasing the field excitation value manually. But what is it to do with this contactor?

The contactor switches the AVR, without power the AVR does not works, if this contactor fails, the AVR fails. On the Other hands, if so, the generator would not work under any conditions, even when the AVRs were changed, much less in the tests performed, and every time the generator was switched on, it would fail, as it would not reach the excitation voltage. However, this was an intermittent failure. Observe the Direct Chain (DC) voltage coil illustrated in **Figure 12**.

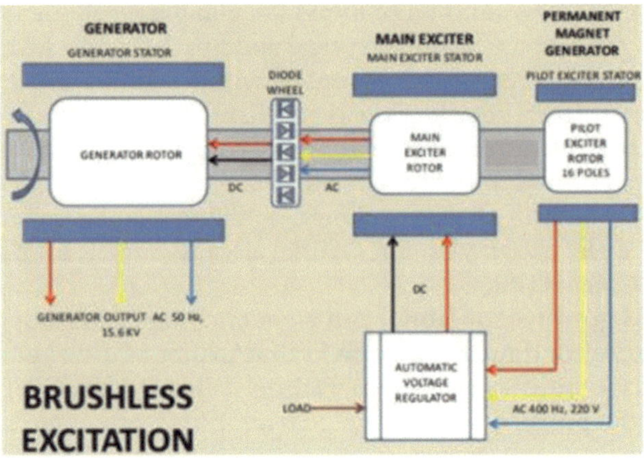

Figure 11.
Excitation system for a brushless alternator [1].

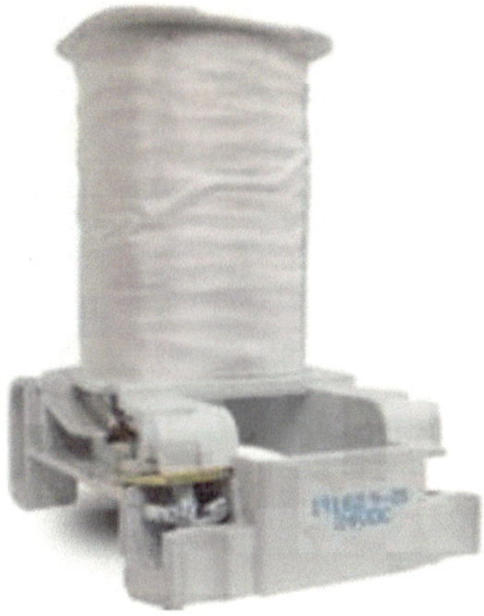

Figure 12.
DC voltage CWM coil [1].

Figure 13.
Disassembled CWM DC contactor [1].

This coil works with a small voltage value, so it must be relatively large to keep the contactor's contacts closed, it must have many turns to have a great magnetomotive force at a little current, so the wire gauge is small. The vessel was more than 10 years and probably that coil was never changed, suffering years of heating on it made the varnish of the wire that constitutes the coil lose its insulation, even more when in use, this short-circuits insulation, causing the coil to lose some turns, decreasing the magnetomotive force. During some heating, the coil loses strength, the ship vibrates, this generates an instant lapse by opening the contacts and closing again, which switches off the AVR, so the generator stays momentarily without voltage control, so there will be no way to work in parallel with another engine. Therefore, it was what happened. **Figure 13** shows the damaged contactor disassembled.

Note that the contacts are in bad conditions. When Chief Engineer was checking the temperature inside of electrical board, he observed the contactor (common for both AVRs) heater than the others. Usually, the number of operations wears the contacts and thus, the contactor has some defect, bad terminals, or something, but it was measured an increase in the temperature of the coil and not in the contacts. in a DC coil, the current is limited by the ohmic resistance of the wire, so it needs a lot of turns, to produce a strong magnetomotive force to close tightly the contacts So, it was a real guess, because it's known that coils are sensitive to temperature, after all the temperature degrades the varnish isolation.

When was disassembled another contactor of a new model, it was possible to compare and ratify the Chief Engineer theory, as illustrated in **Figure 14**.

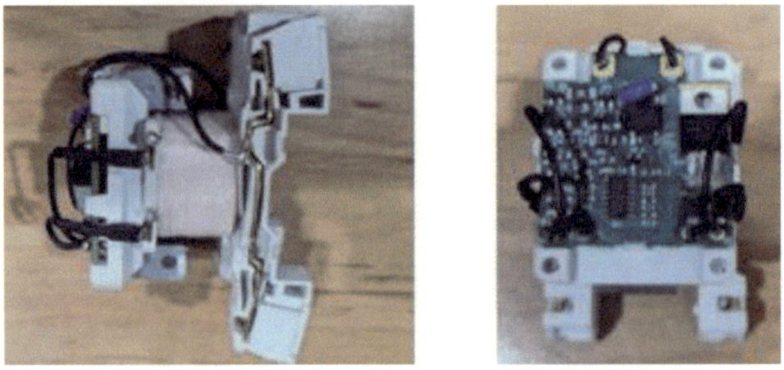

Figure 14.
Disassembled new contactor model [1].

Figure 15.
Timeline for the failures in the generator # 02 synchronization in the busbars.

Notice that the coil of the new models is smaller and has an electronic card, the 24 VDC supply feeds the card which energizes the coil, the card generates an alternating current (AC) wave, which feeds the coil. So, who starts to limit the current value is the impedance of the coil, the number of turns of the coil decreases due to the inductance, so the length of the coil decreases, the ohmic resistance of the wire decreases, thus, the heating decreases, less resistance, less active power dissipated in the form of heat, less heat, greater longevity for the coil lifespan.

Indeed, a simple electric contactor was causing those all problems and compromising the safety navigation of the ship. A better analysis and a correct diagnosis are especially important; otherwise you can create even bigger problems.

3.2 Availability and reliability of generator # 02

Mariners Engineers work in shifts of 08 hours as can be seen in **Figure 15**, showing the timeline of this failure of generator # 02.

MTBF for the failures in generator # 02 in the period of 2 shifts of 08 hours:

MTBF = (2 shiftsx08hours - (1 hours+1.5 hours +1.5 hours))/(3 shutdowns in 16 hours) = (16–4) hours/3 shutdowns = 12/3 = 4 hours.

MTTR is calculated by applying the average time it takes to perform a repair after the failure episode.

MTTR = (Total repair time) / (number of failures). According **Figure 15** we have:

MTTR = (1 + 1.5 + 1.5) / 3 = 4/3 = 1.33 hours.

AVAILABILITY = MTBF/(MTBF+ MTTR).

Therefore the availability of the generator # 02 in 2 shifts is:

AVAILABILITY = 4 hours/(4 hours+1.33 hours) = (4/5.33)x100 = 75%.

The reliability (MTTF) for the DC relay, is considered as for unrepairable devices, in this case it is hope the lifespan of 10 years for this relay, so its failure is consider due to lifespan.

The new relay with AC coil it is expect a longer lifespan than for the DC coil relay due to lesser Joule loss in the AC coil, and Chief Tesla will implement a predictive maintenance plan, measuring the temperature in the electric cubicle in every six month.

4. Conclusion

The objective of this chapter is to contribute with readers responsible for the maintenance of vessels, petro rigs and any industrial processes control plant.

Focused on a smart way for rapidly to solve routine problems on board faced by Marine Engineers, it was prepared a revision in the literature concerning troubleshooting, describing how to prepare and use the Ishikawa diagram and also compare with the dart of Vasconcelos developed by Saulo Vasconcelos. In this real case study the Chief Engineer applied the dart of Vasconcelos to reduce the MTTR and shows the solution for the electric problem of the electric generation # 02, caused by an electric contactor that switches the AVR (Automatic Voltage Regulator), The Mariners Engineers, without troubleshooting the problem replaced three times the AVRs without success to find the root cause of the problem. The Chief Nicola Tesla, applying the dart of Vasconcelos, reached the bull's eye of the problem, the old DC electric contactor. The dart of Vasconcelos directs for root cause and also to avoid future problem, recording the timeline of the problems until the final action, in this real case study the installation of new electric contactor with AC coil with longer lifespan for its AC coil, increasing availability and reliability of the Diesel-electric system of this ship.

Acknowledgements

The authors would like to thanks the LORD GOD for this opportunity, the Admiral Braz de Aguiar Maritime Academy for the authorization to public this work, and to inTechOpen Access Publisher for the virtuous circle created to share knowledge between readers and authors.

Author details

Saulo Vasconcelos and Paulo Vasconcelos*
Admiral Braz de Aguiar Maritime Academy, Belem, Brazil

*Address all correspondence to: douglasic@uol.com.br

Reference

[1] Vasconcelos D, Saulo.
Troubleshooting for Marine Engineers
with Real Cases. Marine Insight
Publisher; 2021. ISBN: 978-65-00-22161-9

Chapter 6

Human Resource Management in National Shipping

Prasadja Ricardianto and Imam Sonny

Abstract

This chapter of this book shows that the mastery of some modern techniques is needed by the ship's crew who work in a management to build a modern ship, in accordance with the theme of this project, namely modern ship engineering, design and operations. The main problem is how the management leaders of this modern ship project can apply transformational leadership style through work-life balance and employee engagement to the ship's crew to make their work effectiveness improved. In carrying out the ship building which takes a quite long time, do the employees can be mutually engaged and can pay equal attention to the work in this project and the work at home? This study uses quantitative method with Structural Equation Modeling. This modern ship building uses some terminologies or variables of human resources such as transformational leadership style, work-life balance, and employee engagement which, based on the research done, directly and positively affect employee's work effectiveness. The key findings of this study indicate that there is a specific model on employee engagement and crew work effectiveness that is very appropriate to be applied to modern ships related to engineering, design and operations.

Keywords: employee engagement, transformational leadership style, ship's crew, work effectiveness, work-life balance, shipbuilding management, modern ships, engineering, design, operations

1. Introduction

According to the data from the Human Resources Division of an Indonesia's national shipping company having some ships berthing in the Port of Tanjung Priok, Jakarta, Indonesia, the employee performance appraisal of the shipping company has a fairly good score in general. There has been an appraisal of one of the ship's officers, who was given a bad score. The employee engagement in general lies on the ship's crew loyalty which is not so strong. In particular, their welfare is not so good as the other similar companies.

The problems of some ships berthing in the Port of Tanjung Priok related to the work effectiveness of ship's crew based on direct interview are such as; (1) Ship installation repair which is not on time, (2) Late and inaccurate reporting on accidents and emergency conditions, (3) The use of personal protective equipment not in accordance with the Standard Operating Procedure, (4) Inaccurate report on shipment, (5) Low understand of International Safety Management Code, and (6) periodic mechanical and electrical maintenance of the ship's hull which is unscheduled.

Transformational leadership style, work-life balance, employee engagement and work effectiveness have been studied in some previous researches. With limited resources the ship's officers should be able to apply an effective approach in order to improve the work effectiveness of their staff. According to [1], in the human resource management research, three terminologies are used, namely transformational leadership, work-life balance and employee engagement. In fact, the variable of work-life balance has a higher position than other variables.

In addition [2], says that a further research will provide understanding of how engagement will prioritize the employee's interests. Human Resource Management is very necessary in the Shipping Industry [3]. In a modern ship building, the term work-life balance as a whole is rated positively by the respondents [4, 5]. Several things to cope with in the variable of work-life balance, especially with ship's crew, are home sickness, fatigue, family problems, discrimination, bad onboard communication during the voyage and bad work relationship [6, 7]. In addition, work engagement is much needed in the management of human resources in the modern ship industry, especially on cruise ships [8, 9].

In general, ship performance improvement in real time can be done through monitoring, analyzing and displaying the ship's performance during service [10]. The recent design of modern ship can be illustrated as an integrated system of its role, operation, and services and using computer-based appliances and based on the design model integrated with the environment [11]. The integrated use of appliances for a new ship design can lead to the quick implementation of ship virtual prototype that has been planned in the early stage. Currently, there is a need for an integrated ship design, through an innovative approach with a new generation of computers [12]. The management of modern ship should be able to adapt it. According to [13], modern ship building should also see the technology integration in management and ship operation, which especially has a close relationship between the ship and the onshore management unit. The development of information and communication technology on smart cruise ships will add to the cruise experience [14]. Technically a modern ship with improved sensor systems using speed log information, GPS, ocean currents and ship dynamics, after weather matching and filtering processes, can increase the average sailing speed [15].

This study tries to get new findings based on hypothetical test whether there are differences in the use of research method, dimensions and indicators and whether there is a direct or indirect relationship. This study tries to find some novelties by comparing the results of this research to the previous ones. The aim of studying the activities of such a modern ship, in accordance with the main theme of this study, is to provide inputs and considerations to the technically modern ship management to pay more attention to some important terminologies for being able to improve their employees' work effectiveness.

1.1 Literature review and hypothesis

1.1.1 Work effectiveness

Effectiveness and efficiency are of performance measurements that can be used to evaluate the employee performance [16]. The management of a shipping business is expected to be able to make appropriate decisions, as a main power to enhance the safety of their employees. It is necessary to improve the effectiveness of the ship's crew of national shipping companies to be able to compete with the private similar companies. Work effectiveness according to theory [17], has several research aspects; such as quantity, quality, reliability, attendance, and ability to collaborate. The variable of work effectiveness has several research dimensions, namely; 1) Work quality, 2) Work quantity, 3) Punctuality, 4) Work effectiveness, and 5) Independent attitude.

Conceptually, it can be concluded that work effectiveness is the output of employee's work consisting of work quality, work quantity, punctuality and employee job satisfaction used to achieve the goals. The aspects of work effectiveness are; 1) Punctuality, 2) Work quantity, 3) Work quality, and 4) Work suitability.

1.1.2 Transformational leadership style

According to [18, 19], transformational leadership style is an ability to bring significant changes to the employees in an organization. A leadership style with democratic dimension, according to [20], can give freedom to employees and involve them in the decision making, which is more productive. In addition, the results of the researches done by [12, 13], also indicate the significant influence of transformational leadership on the trust in a leader, and, in turn, it gives a positive impact on the employee engagement. A leader has orientation to his staff and their work. Moreover [21], explains that transformational leadership style can potentially has an influence in improving the performance and quality of ship's crew. Besides that, democratic leadership style is also very effective although further studies show various results. Transformational leadership behavior promotes teamwork, high performance expectations, and individual consideration significantly [22, 23]. Based on the opinion of [24], the characteristics of transformational leadership style, by developing several aspects, are; 1) Ideal influence; 2) Inspirational Motivation; 3) Intellectual Stimulation; and 4) Individual consideration. Ideal influence is the most important aspect of transformational leadership.

Conceptually, based on the theoretical studies and previous relevant researches from some experts who support this study, it can be concluded that transformational leadership style is the way of a leader to design, influence, and ask for the engagement of his subordinates to achieve the organizational goals consisting of several aspects, such as; 1) ideal influence or charisma, 2) inspirational motivation, 3) intellectual stimulation, 4) individual consideration, and 5) democratic style.

1.1.3 Work-life balance

According to some experts like [25, 26], Work-life balance is guided by the working hours and positive results when someone is able to combine work role and family role. In the opinion of [27–29] Work-life balance has become an important variable in coping with one of employee management problems. There is a significant positive relationship between family life and work. Whereas Work-life balance according to [30] closely related to some other aspects like the job characteristics, allowances, support from superiors, employee's working culture, job satisfaction.

According to Frone, Yardley and Markel in [31], Work-life balance has five aspects, namely; (1) Time management; (2) Social life outside the work; (3) Work life balance; (4) Work-family balance; and (5) Ability to keep involved in the non-work interests and activities. Work-life balance is also related to job demand, intention to move and psychological tension, but in positive ways it can involve family and result in job satisfaction [32, 33]. These five aspects are based on the theory and some previous researches that conceptually support this study to become the main aspect in the terminology of Work-life balance.

1.1.4 Employee engagement

An employee who wants to be engaged in the company must have commitment to the job and working motivation to achieve a high performance [19, 34, 35]. In the previous research, employee engagement according to [36–38] is a condition which is expected by the company, having psychological bond, having organizational goals and showing bond, commitment, vigor and focus on work. According to Khan, an employees will be tied to their work and when this happens they will totally work; physically, cognitively and emotionally [39]. Aon Hewitt, who develops a model of employee engagement (**Figure 1**), explains that the five terminologies of Human Resources Management are established to be a Grand Theory and variable unity in this study.

Employee engagement is something satisfactory related to vigor, dedication, and absorption which is significantly and positively related to the performance of organization [41–44]. Whereas [45], proposes three aspects of employee engagement, namely Say, Stay and Strive. Strong employee engagement will be able to influence a work [46, 47]. In the opinion of [48] the engagement strategy implemented by organizations has reached a satisfactory level. Moreover [49, 50], state that an organization may have a competitive advantage in the form of increasing employee engagement and can affect the welfare. A research carried out by Schaufeli of Utrecht Work Engagement Scale (UWES) uses three dimensions; vigor, dedication and absorptions [51]. The results of correlation and regression from the research of [52], show that team support is the strongest predictor of an engagement.

Conceptually, based on the theory and previous researches from some experts who support this study, it can be concluded that employee engagement is a positive attitude of an employee to the engagement, commitment to working, enthusiasm in working, comfort in working with some aspects, namely: 1) Vigor, 2) Dedication, 3) Absorption, and 4) Stay.

1.1.5 Hypothesis

H1. Transformational Leadership Style has influences on Employee Engagement
H2. Work-Life Balance has influences on Employee Engagement

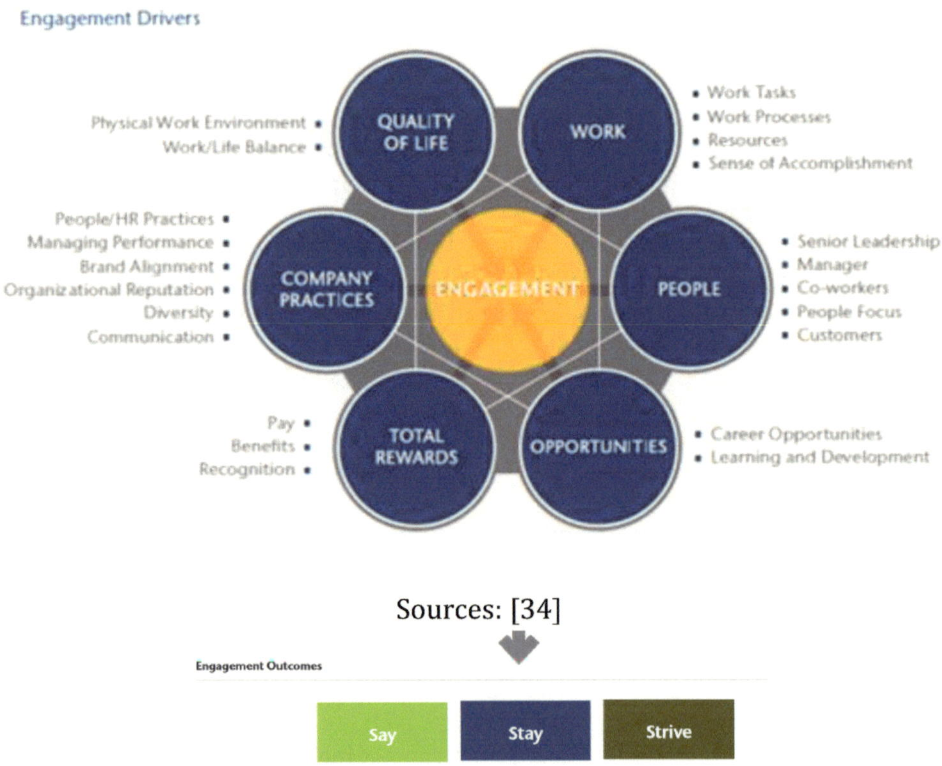

Figure 1.
Model: Engagement drivers. Sources: [40].

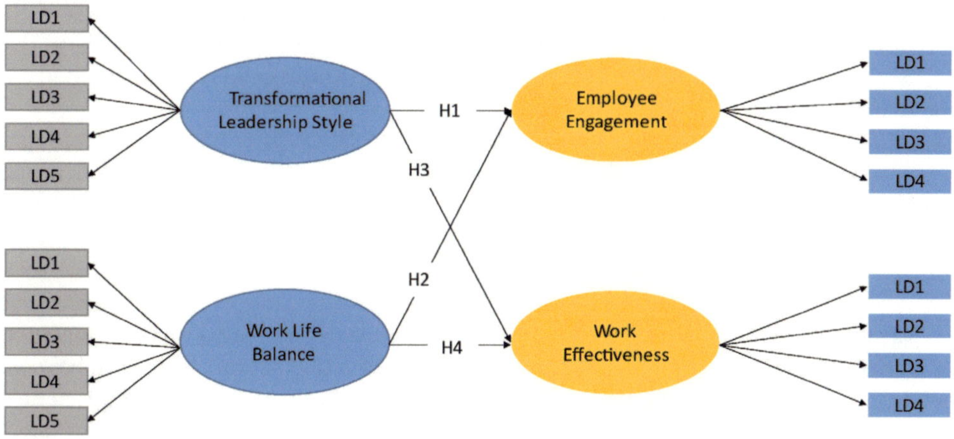

Figure 2.
Conceptual framework.

H3. Transformational Leadership Style has influences on Work Effectiveness
H4. Work-Life Balance has influences on Work Effectiveness
H5. Employee Engagement has influences on Work Effectiveness

Some terminologies in the Human Resource Management such as; Transformational Leadership Style, Work-Life Balance, dan Employee Engagement are important variables to improve the work effectiveness of ship's crew.

Below is the conceptual framework of this research (**Figure 2**).

2. Research method

This research has four main variables in this case, such as, transformational leadership style, work life balance, employee engagement and work effectiveness. Each variable studied includes several dimensions and indicators. The analytical tool in this quantitative study uses Structural Equation Modeling (SEM) with the help of the Lisrel 8.7.1 program. Hair et al. [53] explained that the use of SEM allows simultaneous analysis of a series of relationships, thus providing statistical efficiency. The use of Structural Equation Modeling is also used to test research hypotheses.

The sampling technique in this study is called Cluster Systematic Sampling. The Slovin formula was used to obtain a sample of 290 crew members from a total population of 1,050 crew members from seven modern passenger ships with a capacity of 2000 passengers on each ship. The results of this research questionnaire are considered reliable and valid in previous trials on 30 respondents from crew members taken from one of the seven modern ships. The validity test on the work effectiveness, transformational leadership style, work-life balance and employee engagement instruments resulted in most of the items being valid statements. The results of the reliability test for all variables show that they are above 0.9 and it means that they have high reliability.

The use of the Lisrel analysis tool from Structural Equation Modeling is currently being carried out by several previous researchers to test the conceptual framework.

Previous research that has a positive effect on work effectiveness using SEM was carried out by [54]. Several studies on transformational leadership styles that have used SEM were also conducted by [48, 49, 55, 56]. The study of work-life balance also applies a lot of SEM models adapted by [57–59]. Meanwhile, other studies related to employee engagement that use SEM are described by [60, 61]. Several researchers have previously been recorded using the SEM model with the help of the Lisrel program [62–64].

3. Discussion

3.1 Result of model test

Based on the test result of Confirmatory Factor Analysis (CFA) on the construct of transformational leadership style, that all the indicators in the dimension have been valid. Likewise, the value of CR is 0.96 (> 0.70) dan VE is 0.82 (> 0.50). The test result of CFA on the construct of work of balance that all the indicators in the dimension have been valid. Likewise, the value of CR is 0.97 (> 0.70) dan VE is 0.87 (> 0.50). The test result of CFA on the construct of employee engagement that all the indicators in the dimension have been valid. Likewise, the value of CR is 0.98 (> 0.70) dan VE is 0.91 (> 0.50). Whereas, the test result of CFA on the construct of work effectiveness that all the indicators in the dimension have been valid . Likewise, the value of CR is 0.96 (> 0.70) dan VE is 0.87 (> 0.50).

Based on the test result of Confirmatory Factor Analysis (CFA) on the construct of transformational leadership style, work life balance, employee engagement and work effectiveness from all the summarized indices, can be concluded that all the indicators used in this study have good validity and reliability values. Thus, it can be illustrated as the result of the model test (**Figure 3**).

3.1.1 Hypothetical test

Some of the results of hypothesis testing are described in the following tables (**Figure 3**).

Based on **Table 1**, transformational leadership style and work-life balance affect employee engagement, so that both hypotheses 1 and 2 are accepted.

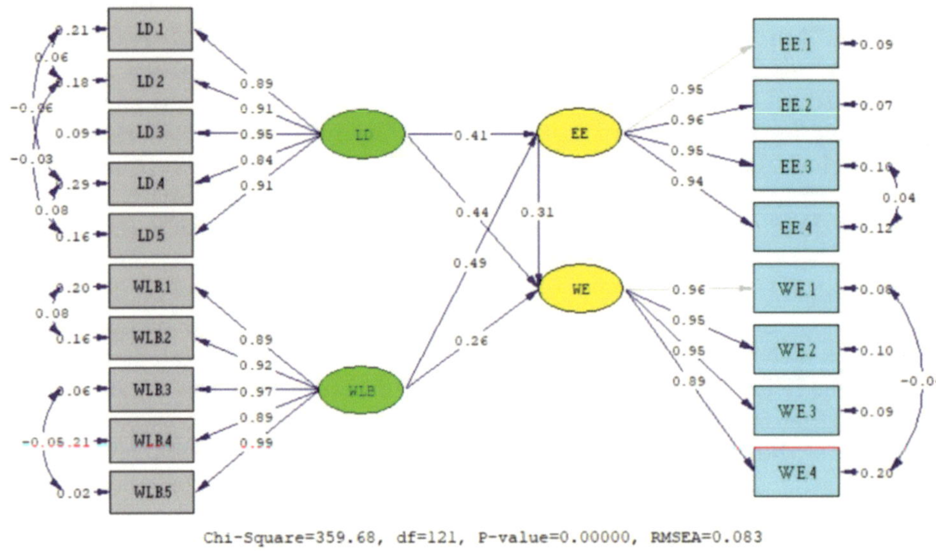

Figure 3.
Research model.

Path	Coefficient	T_{Value}	Results
Transformational Leadership Style → Employee Engagement	0.41	7.65	Influential
Work Life Balance → Employee Engagement	0.49	8.90	Influential
Structural Equation EE = 0.41LD + 0.49WLB + Ɛ			

Table 1.
The influence of transformational leadership style and work life balance on employee engagement.

H1: The influence of transformational leadership style on employee engagement.

Previous relevant study results state that transformational leadership style directly and positively affects employee engagement. The theoretical studies state that leadership style supports employee engagement. For example, the theoretical opinion of [65] states that leadership style supports engagement. Several previous studies by [66, 67], it has been studied that employee engagement can become a mediation for transformational leadership and company branding. Some studies say that transformational leadership style has a positive and direct influence on employee engagement [68, 69].

The result of another research shows that the relationship between the level of employee engagement and the influence of leadership can be measured through the level of management commitment [70]. Another leadership style, i.e. democratic style, shows its quite strong relationship with employee engagement [71]. Another research by [72] states that transformational leadership style affects the employee satisfaction of an Indonesian company through employee engagement. Good transformational leadership will improve employee engagement. However [73], in their study find the indirect relationship between transformational leadership and engagement.

Thus, the results of previous relevant studies support this study. It means leadership style has a direct and positive influence on employee engagement.

H2: The influence of work-life balance on employee engagement.

The previous relevant research by [31] test the hypothesis resulting in that work-life balance is necessary for employee engagement. Employee engagement is increasingly being seen as a win-win strategy for companies, employees, and their communities. Moreover [51], states that work-life balance is getting more important for employee engagement. Another research says that the policy and practice of work-life balance will get benefits from higher employee engagement [52, 53]. In another study through a comparison of indirect effects, the findings of [74] show that employee engagement and work-life balance can act as mediators. Moreover [75], explains that organizations support work-life balance and employee engagement. Another research by [72] explains that work-life balance affects employee satisfaction in an Indonesian company through employee engagement. The result of another research by [76] in Vietnam, shows that work-life balance and work stress positively affects employee engagement.

Thus, the results of this study are in accordance with previous theoretical studies and related research. This means that work-life balance has a positive and positive effect directly on employee engagement.

From **Table 2**, shows that transformational leadership style and work-life balance affect employee engagement, so that both hypotheses 3 and 4 are accepted.

H3: The influence of transformational leadership style on work effectiveness.

Previous relevant study results state that transformational leadership style directly and positively affects work effectiveness. The theoretical studies state that transformational leadership style influence work effectiveness. The findings of

Path	Coefficient	T_{Value}	Results
Transformational Leadership Style → Work Effectiveness	0.44	9.43	Influential
Work-Life Balance → Work Effectiveness	0.26	5.82	Influential
Structural Equation WE = 0.44LD + 0.36WLB + Ɛ			

Table 2.
The influence of transformational leadership style and work-life balance on work effectiveness.

researches [77], confirm that performance improvement is highly affected by many factors; one of them is leadership. This study is also in line with the opinion [78], stating that the variable of leadership directly affects employee's work effectiveness. Based on the study [79] on the business of shipyard in Indonesia, transformational leadership positively and significantly affects employee performance. The result of another research shows that transformational leadership style directly and positively affects work effectiveness [80]. Moreover [81], states that transformational leadership also develops a research model which is designed to assess the effectiveness and performance of leadership.

In conclusion, the results of this study are in line with previous theoretical studies and related research. Then, transformational leadership style directly and positively affects work effectiveness.

H4: The influence of work-life balance on work effectiveness.

Previous relevant study results state that work-life balance directly and positively affects work effectiveness. The theoretical studies state that work-life balance supports work effectiveness. The researches [80, 82] prove that better work-life balance leads to the improvement of employee performance. The arrangement of work-life balance can be related to the perception of individual performance, that those who have better performance are considered as able to take advantage of work-life balance.

In conclusion, the results of this study are in line with previous theoretical studies and related research. Then, work-life balance directly and positively affects work effectiveness.

From **Table 3** shows that Employee engagement has an effect on work effectiveness, the end result is hypothesis 5 is accepted.

H5: The influence of employee engagement on work effectiveness.

The results of previous relevant researches state that employee engagement directly and positively affects work effectiveness. The theoretical studies state that employee engagement supports work effectiveness. Theoretically [61, 83, 84], state that employee engagement has an impact on the improvement of performance. The result of another research [85], also reveals the positive influence of perceived organizational support on the employee performance mediated by employee engagement. Another researcher shows the way organizations can enhance the engagement among their employees which will improve the effectiveness of organizations [86]. Finally, the need for strong organizational commitment and high work engagement is the determining factor of success to achieve higher performance [37, 87].

Path	Coefficient	T_{Value}	Results
Employee Engagement → Work Effectiveness	0.31	6.45	Influential
Structural Equation WE = 0.31EE + Ɛ			

Table 3.
The influence of employee engagement on work effectiveness.

In conclusion, the results of this study are in line with previous theoretical studies and related research. Then, employee engagement directly and positively affects work effectiveness.

4. Mediation effect test

This research also conducts a testing related to intervening or mediating variable, so it is necessary will be able to state whether employee engagement is appropriate to produce the intervening variable between transformational leadership style and work-life balance as the independent variables and work effectiveness as the dependent variable.

Employee engagement as an intervening variable between transformational leadership style and work-life balance as well as work effectiveness which explains that initially there is a significant direct relationship. For example, the value of t for the direct influence of transformational leadership style on the work effectiveness is 7.65 and after the inclusion of employee engagement the value of t decreases to 5.09 and still significant with $t_{-value} > 1.96$ (**Table 4**).

The ways to improve the work effectiveness of modern ship's in line with the theme of this project, i.e. Modern Ship Engineering, Design and Operations, are through appropriate analysis, reducing fuel consumption through optimum ship's trim, on time machine setting or propeller cleaning [10]. In terms of engineering and operation, the ship's crew must pay attention to modern ship management such as the data of hindcast wind and the data of wind sensor, speed log sensor, and the filtering process, in order to get an accuracy in the daily report, the number of sensors and the Data Acquisition Systems (DAQs) where each of them can be seen as the node of data collection sources [15, 88].

According to [78, 79], in their researches, transformational leadership style and employee engagement, significantly affect the employee work effectiveness. In the time of Covid-19 pandemic, how the leadership can be responsible for supporting the tourism sector through the management of modern ship in order to be able to manage the crisis [89]. Radic and Radic et al. [9, 90] in their research state that there is an opportunity in a modern ship to enhance the employee engagement by enhancing the factors related to communication and advancement, the capacity to be engaged and the behavior of engagement. Study by [91], asserts the existence of significant and positive engagement related to the degree of retention, and some contributing factors. According to [8], the crew work in a bad condition and it determines their engagement, contributing to the identification of the dimensions of work quality that need to be improved by the Human Resources Manager in modern ships.

Work-life balance in modern ships must pay attention to the significant predictor of welfare such as the employee's multicultural environment [4]. Especially for shipping industry, seafarers who work on modern ships like a cruise ship, Barnett

Influence	Direct		Indirect		Results
	Coef	t_{-value}	Coef	t_{-value}	
Transformational Leadership Style → Work Effectiveness	0.44		9.43		Influential
Work Life Balance → Work Effectiveness	0.26		5.82		Influential
Structural Equation $WE = 0.44LD + 0.36WLB + \varepsilon$					

Table 4.
Single mediation test through employee engagement.

explains the tendency of complaining the work condition which is full of stress and negatively affecting their work-life balance in the long term [3]. Through a quantitative approach [5], explains in his theory that work-life balance is highly necessary for the crew of modern ship.

Some studies are in line with this study that ship management through engineering, Smart Maritime Ecosystem, digital features of modern ship, the utilization of Communication Information Technology, especially in terms of efficiency, sustainability, security, and experience personalization, as well as challenges and limitations related to the improvement of that technology [14]. Improvements in several human resource variables through research on seven modern ships at the Tanjung Priok port, Jakarta, Indonesia that are well integrated will be able to improve engineering, design and operational fields such as the Smart Maritime Ecosystem, digital features of modern ships, utilization of information and communication technology, systems sensors, GPS.

5. Conclusion and recommendation

Then it can be concluded that the effectiveness of work in the national shipping company is directly affected by the transformational leadership style of the ship's officers, work-life balance, and ship's crew engagement. Based on the results of research on modern ship building, the novelty of the research lies in several aspects of human resources. On the effectiveness of the work of the crew at the national shipping company, the novelty in the aspect of work reliability.

The novelty of the transformational leadership style of research lies in the aspect of intellectual stimulation. In the work-life balance variable lies in the work-family balance aspect, while the engagement of the crew is seen as new in the aspect of loyalty. The main novelty in this study is that the employee engagement variable is a mediating variable which is expected to be a key assessment aspect, especially for shipping employees on modern ships.

Overall, on the work effectiveness, It is hoped that the crew can immediately analyze several aspects of non-conformity in work, work safety and hazardous conditions immediately, correctly and accurately. This study is potential to support national shipping companies, particularly modern ship companies to improve the work effectiveness of employees, especially ship's crew.

The key finding of this research which is also a novelty of this research is that through variable of transformational leadership, work-life balance and employee engagement will be able to improve ship crew's work effectiveness. Thus, this modern ship research is integrated in line and supports previous research, especially the relationship between the four variables of human resources with the fields of engineering, design and operational. This research is a model that will improve the quality of modern shipbuilding with large weight and high speed.

Acknowledgements

Writer team hereby give thanks the editors, Proofreader and the Directorate of Research and Community Service the Ministry of Research and Technology and Higher Education of the Republic of Indonesia as the financial supporter. Ship's Officers and Ship's Crew especially the ships berthing in the Port of Tanjung Priok, Jakarta. For all respondents we say thanks, also Trisakti Institute Transportation and Logistics that has assigned the authors to the Port of Tanjung Priok and for all assistance provided to accomplish this study.

Author details

Prasadja Ricardianto* and Imam Sonny
Trisakti Institute of Transportation and Logistics, Jakarta, Indonesia

*Address all correspondence to: ricardianto@gmail.com

References

[1] N. Garg, "Employee engagement and individual differences: A study in Indian context," Manag. Stud. Econ. Syst., vol. 1, no. 1, pp. 41-50., 2014, DOI:10.2139/ssrn.2693436.

[2] J. R. Halbesleben, "Individual-level outcomes of employee engagement: a conservation of resources framework," in A Research Agenda for Employee Engagement in a Changing World of Work, 2021, pp. 87-100.

[3] L. D. Caesar and J. Fei, "Work-life balance," in Managing Human Resources in the Shipping Industry, J Fei (ed)., 1st Eds., Ed. Routledge, 2018, pp. 107-128.

[4] C. Lütkenhaus, "Work–life balance on board cruise ships: A survey on crew members' well-being," Tour. Mar. Environ., vol. 15, no. 2, pp. 73-83., 2020, DOI:10.3727/154427320X15833169859482.

[5] C. Schädel, Work-Life Balance among Cruise Ship Crews: A Quantitative Research Approach. Hamburg: Bachelor+ Master Publication, 2011.

[6] K. Hoffmann-Burdzińska and M. Rutkowska, "Work life balance as a factor influencing well-being.," J. Posit. Manag., vol. 6, no. 4, 2015, DOI:10.12775/JPM.2015.024.

[7] A. R. G. Manalo, N. R. Mercado, D. F. Paragas, J. C. C. Tenorio, and J. C. Dotimas, "The challenges of Filipino seafarers onboard: Basis for work life balance.," LPU-Laguna J. Int. Tour. Hosp. Manag., vol. 3, no. 1, pp. 157-184., 2015.

[8] A. Ariza-Montes, A. Radic, J. M. Arjona-Fuentes, H. Han, and R. Law, "Job quality and work engagement in the cruise industry," Asia Pacific J. Tour. Res., vol. 26, no. 5, pp. 469-487., 2021, DOI:10.1080/10941665.2020.1866625.

[9] A. Radic, "All hands on deck: Employee engagement on cruise ships," African J. Bus. Manag., vol. 11, no. 6, pp. 117-126., 2017.

[10] I. Van den Boom, H., Van der Hout and M. Flikkema, "Speed-Power Performance of Ships during Trials and in Service," 2008.

[11] T. K. Jauhari, A. Maimun, and C. L. Siow, "Review of Systems Engineering Methods, Techniques and Tools for Ship Design as Large and Complex Systems," 2019.

[12] U. la Monaca, S. Bertagna, A. Marinò, and V. Bucci, "Integrated ship design: An innovative methodological approach enabled by new generation computer tools," Int. J. Interact. Des. Manuf., vol. 14, no. 1, pp. 59-76., 2020.

[13] B. Suresh, "Challenges and Potential of Technology Integration In Modern Ship Management Practices," 2013.

[14] M. Vafeidou, "Smart Cruise Ships: in What Way Information and Communication Technologies Are Revolutionizing the Cruise Experience.," International Hellenic University Scholar Works, 2019.

[15] J. Park, B. Kim, J. Jeong, H. Park, D. Jeong, and K. Ahn, "A Comparative Analysis of Ship Speed-Power Performance Based on the Noon Reports and Recorded Sensor Data: Overcoming Sensor Issues," 2017.

[16] I. Bartuševičienė and E. Šakalytė, "Organizational assessment: Effectiveness vs. efficiency.," Soc. Transform. Contemp. Soc., vol. 1, pp. 45-53., 2013.

[17] R. L. Mathis and J. H. Jackson, Human resource management, 13th Eds. USA: South-Western: Cengage Learning, 2010.

[18] R. Daft, The leadership experience, 6th Eds. USA: Cengage Learning, 2015.

[19] S. P. Robbins and T. A. Judge, Organizational Behavior. USA: Pearson, 2015.

[20] N. Sad and D. Bosiok, "Leadership styles and creativity," Online J. Appl. Knowl. Manag., vol. 1, no. 2, pp. 64-77, 2013.

[21] N. Xiang, "Transformational leadership and its implication for leadership at sea," Int. Assoc. Marit. Univ. AGA11, pp. 121-131, 2019.

[22] N. Callow, M. J. Smith, L. Hardy, C. A. Arthur, and J. Hardy, "Measurement of transformational leadership and its relationship with team cohesion and performance level," J. Appl. Sport Psychol., vol. 21, no. 4, pp. 395-412., 2009, DOI:10.1080/10413200903204754.

[23] H. Li, N. Sajjad, Q. Wang, A. Muhammad Ali, Z. Khaqan, and S. Amina, "Influence of transformational leadership on employees' innovative work behavior in sustainable organizations: Test of mediation and moderation processes," Sustainability, vol. 11, no. 6, p. 1594., 2019, DOI:10.3390/su11061594.

[24] M. Tschannen-Moran, "Fostering organizational citizenship in schools: Transformational leadership and trust," in Journal of Educational Administration., 2003, pp. 1-36.

[25] J. L. Gibson, J. M. Ivancevich, H. Donnely Jr, and R. Konopaske, Organization: Behaviour, structure, and process, Fourteenth. Boston: Boston: McGraw-Hill, 2012.

[26] L. J. Mullins, Management and Organizational Behavior. Pearson Education Limited, 2013.

[27] Clark, M. D. Hartline, and K. C. Jones, "The effects of leadership style on hotel employees' commitment to service quality," Cornell Hosp. Q., vol. 50, no. 2, pp. 209-231., 2009.

[28] D. E. Guest, "Perspectives on the study of work-life balance," Soc. Sci. Inf., vol. 41, no. 2, pp. 261-266., 2003.

[29] M. Deery and L. Jago, "Revisiting talent management, work-life balance and retention strategies," Int. J. Contemp. Hosp. Manag., vol. 27, no. 3, pp. 453-472, 2015, DOI:10.1108/IJCHM-12-2013-0538.

[30] R. Baral and S. Bhargava, "Work-family enrichment as a mediator between organizational interventions for work-life balance and job outcomes," J. Manag. Psychol., vol. 25, no. 3, pp. 274-300, 2010, DOI:10.1108/0268394 1011023749.

[31] L. P. Parkes and P. H. Langford, "Work-life balance or work-life alignment? A test of the importance of work-life balance for employee engagement and intention to stay in organizations," J. Manag. Organ., vol. 14, 2008, DOI:10.5172/jmo.837.14. 3.267.

[32] T. Kalliath and P. Brough, "Work life balance: A review of the meaning of the balance construct," J. Manag. Organ., vol. 14, no. 3, pp. 323-327, 2008, DOI:10.1017/S1833367200003308.

[33] P. Brough et al., "Work–life balance: A longitudinal evaluation of a new measure across Australia and New Zealand workers," Int. J. Hum. Resour. Manag., vol. 25, no. 19, pp. 2724-2744, 2014, DOI:10.1080/09585192.20 14.899262.

[34] M. Armstrong and S. Taylor, Armstrong's Handbook of Reward Management Practice: Improving Performance through Reward. London and Philadelphia: Kogan Page Limited, 2014.

[35] R. Kreitner and A. Kinicki, Organizational Behavior, 9th Eds. New York: McGraw-Hill Irwin, 2011.

[36] W. H. Macey and B. Schneider, “The meaning of employee engagement,” Ind. Organ. Psychol., vol. 1, no. 1, pp. 3-30., 2008.

[37] S. Markos and M. S. Sridevi, “Employee engagement: The key to improving performance,” Int. J. Bus. Manag., vol. 5, no. 12, p. 89, 2010, DOI:10.5539/ijbm.v5n12p89.

[38] A. Mazzei, “Employee engagement,” Int. Encycl. Strateg. Commun., pp. 1-6, 2018, DOI:10.1002/9781119010722.iesc0068.

[39] A. M. Saks, “Antecedents and consequences of employee engagement,” J. Manag. Psychol., vol. 21, no. 7, pp. 600-619, 2006, DOI:10.1108/02683940610690169.

[40] AON Hewitt, “Trends in Global Employee Engagement,” 2012. [Online]. Available: https://www.aon.com/attachments/human-capital-consulting/2012_TrendsInGlobalEngagement_Final_v11.pdf.

[41] W. B. Schaufeli and A. B. Baker, “UWES Utrecht Work Engagement Scale Preliminary Manual,” utrecht, 2004. [Online]. Available: http://www.wilmarschaufeli.nl/publications/Schaufeli/Test Manuals/Test_manual_UWES_English.pdf.

[42] I. Robertson, A. J. Birch, and C. L. Cooper, “Job and work attitudes, engagement and employee performance where does psychological well-being fit in leadership,” Leadersh. Organ. Dev. J., vol. 33, pp. 224-232., 2012, DOI:10.1108/01437731211216443.

[43] C. Timms et al., “Positive pathways to engaging workers: Work–family enrichment as a predictor of work engagement,” Asia Pacific J. Hum. Resour., vol. 53, pp. 490-510., 2015, DOI:10.1111/1744-7941.12066.

[44] D. E. Frederick and T. J. Vander Weele, “Meta-analysis of job crafting shows positive association with work engagement.,” Cogent Psychol., vol. 7, no. 1, p. 1746733, 2020, DOI:10.1080/23311908.2020.1746733.

[45] R. S. Wellins, P. Benthal, and M. Phelps, “Employee Engagement: The Key To Realizing Competitive Advantage,” 2005. [Online]. Available: https://www.ddiworld.com/ddi/media/monographs/employeeengagement_mg_ddi.pdf?ext=.pdf.

[46] R. Monica and R. Krishnaveni, “Enablers of employee engagement and its subsequent impact on job satisfaction.,” Int. J. Hum. Resour. Dev. Manag., vol. 18, no. 1-2, pp. 5-31., 2018, DOI:10.1504/IJHRDM.2018.092290.

[47] L. Zhang, Y. Zhang, M. Dallas, S. Xu, and J. Hu, “How perceived empowerment HR practices influence work engagement in social enterprises – A moderated mediation model,” Int. J. Hum. Resour. Manag., 2018, DOI:10.1080/09585192.2018.1479874.

[48] M. Kaliannan and S. N. Adjovu, “Effective employee engagement and organizational success: A case study,” Procedia-social Behav. Sci., vol. 172, pp. 161-168., 2015.

[49] N. Adnan, O. K. Bhatti, and W. Farooq, “Relating ethical leadership with work engagement: How workplace spirituality mediates?,” Cogent Bus. Manag., vol. 7, no. 1, p. 1739494, 2020.

[50] A. Rahman, P. Björk, and A. Ravald, “Exploring the effects of service provider’s organizational support and empowerment on employee engagement and well-being,” Cogent Bus. Manag., vol. 7, no. 1, pp. 1-19, 2020, DOI:10.1080/23311975.2020.1767329.

[51] W. B. Schaufeli, M. Salanova, V. González-Romá, and A. B. Bakker, "The measurement of engagement and burnout: A two sample confirmatory factor analytic approach," J. Happiness Stud., vol. 3, no. 1, pp. 71-92., 2002, DOI:10.1023/A:1015630930326.

[52] J. Xu and H. C. Thomas, "How can leaders achieve high employee engagement?," Leadersh. Organ. Dev. J., vol. 32, no. 4, pp. 399-416, 2011, DOI:10.1108/01437731111134661.

[53] J. F. Hair, W. C. Black, B. J. Babin, and R. E. Anderson, Multivariate Data Analysis, 7th Eds. Pearson Education Limited, 2014.

[54] Y. N. Naik and M. Bisht, "Structural equation model for effectiveness of technical managers in information technology industry.," Int. J. Supply Chain Manag., vol. 3, no. 3, 2014.

[55] H. Kim, "Work-life balance and employees' performance: The mediating role of affective commitment," glob. Bus. Manag. Res. An Int. J., vol. 6, no. 1, p. 38, 2014.

[56] A. K. Shrestha, "Leadership styles, subordinates' Satisfaction with the Leader and Perceived Effectiveness.," 5, 2012.

[57] T. Tavassoli and A. Sune, "A national study on the antecedents and outcomes of work-life balance in Iran.," PEOPLE Int. J. Soc. Sci., vol. 3, no. 3, 2018.

[58] R. A. Banu, "A structural equation model: II for work-life balance of IT professionals in Chennai.," Serbian J. Manag., vol. 11, no. 1, pp. 141-148, 2016.

[59] J. M. Haar, M. Russo, A. Suñe, and A. Ollier-Malaterre, "Outcomes of work–life balance on job satisfaction, life satisfaction and mental health: A study across seven cultures.," J. Vocat. Behav., vol. 85, no. 3, pp. 361-373., 2014.

[60] J. Hanaysha, "Testing the effects of employee engagement, work environment, and organizational learning on organizational commitment.," Procedia-social Behav. Sci., vol. 229, pp. 289-297., 2016.

[61] J. Anitha, "Determinants of employee engagement and their impact on employee performance.," Int. J. Product. Perform. Manag., vol. 63, no. 3, pp. 308-323, 2014.

[62] J. Sadeli, "The influence of leadership, talent management, organizational Cultureand organizational support on employee engagement.," Int. Res. J. Bus. Stud., vol. 5, no. 3, 2015.

[63] S. Rani, T. J. Kamalanabhan, and M. Selvarani, "Work/life balance reflections on employee satisfaction. Serbian journal of management," Serbian J. Manag., vol. 6, no. 1, pp. 85-96, 2011.

[64] A. Carmeli, R. Reiter-Palmon, and E. Ziv, "Inclusive leadership and employee involvement in creative tasks in the workplace: The mediating role of psychological safety.," Creat. Res. Journal, vol. 22, no. 3, pp. 250-260, 2010.

[65] M. Armstrong, Armstrong's Handbook of Reward Management Practice: Improving Performance through Reward. London and Philadelphia: Kogan Page Limited., 2010.

[66] S. Sahu, A. Pathardikar, and A. Kumar, "Transformational leadership and turnover: Mediating effects of employee engagement, employer branding, and psychological attachment," Leadersh. Organ. Dev. J., vol. 39, no. 1, pp. 82-99, 2017, DOI:10.1108/LODJ-12-2014-0243.

[67] T. Besieux, E. Baillien, A. L. Verbeke, and M. C. Euwema, "What

goes around comes around: The mediation of corporate social responsibility in the relationship between transformational leadership and employee engagement," Econ. Ind. Democr., vol. 39, no. 2, pp. 249-271., 2018, DOI:10.1177/0143831X15614914.

[68] A. J. Wefald, R. J. Reichard, and S. A. Serrano, "Fitting engagement into a nomological network: The relationship of engagement to leadership and personality.," J. Leadersh. Organ. Stud., vol. 18, no. 4, pp. 522-537., 2011, DOI:10.1177/1548051811404890.

[69] F. Tabak and N. T. Hendy, "Trust as a mediator of the impact of organizational job Embeddedness and perceived organizational support," Organ. Manag. J., vol. 13, no. 1, pp. 21-31, 2016, DOI:10.1080/15416518.2015.1116968.

[70] M. Parimalam and A. Mahadevan, "The relationship between leadership and employee engagement: An employee course perspective," J. Contemp. Res. Manag., vol. 7, no. 4, pp. 27-35, 2012.

[71] L. Yao, F. L. Kee Shin Woan, M. H. B. Ahmad, and G. Kuantan, "The relationship between leadership styles and employee engagement: Evidences from construction companies in Malaysia," Soc. Sci., vol. 12, pp. 984-988., 2017.

[72] S. Erwin, S. T. Y. Rahmat, N. D. Angga, and P. Semerdanta, "Transformational leadership style and work life balance: The effect on employee satisfaction through employee engagement," Russ. J. Agric. Socio-Economic Sci., vol. 91, no. 7, 2019.

[73] A. J. Hawkes, A. Biggs, and E. Hegerty, "Work engagement: Investigating the role of transformational leadership, job resources, and recovery," J. Psychol., vol. 151, no. 6, pp. 509-531., 2017, DOI:10.1080/00223980.2017.1372339.

[74] R. Kaur and G. Randhawa, "Supportive supervisor to curtail turnover intentions: do employee engagement and work–life balance play any role?"," Evidence-based HRM, vol. ahead-of-p, 2020, DOI:10.1108/EBHRM-12-2019-0118.

[75] E. K. Murvin, "The Correlation between Manager Work-Life Balance and Employee Engagement. Walden Dissertations and Doctoral Studies. 6762," Walden University, 2019.

[76] L. Dinh, "Determinants of employee engagement mediated by work-life balance and work stress," Manag. Sci. Lett., vol. 10, no. 4, pp. 923-928., 2020, DOI:10.5267/j.msl.2019.10.003.

[77] F. Arif, "The Influence of Transformational Leadership, Discipline, Productivity On Employee Performance," Sci. J. Reflect. Econ. Accounting, Manag. Business, vol. 4, no. 1, pp. 51-60, 2021, DOI:10.37481/sjr.v4i1.247.

[78] P. E. Hasibuan, M. Akbar, and T. Suyatno, "The effect of leadership, employee engagement, and self learning on effectiveness of Employee's work," Int. J. Sci. Res. Manag., vol. 6, no. 7, pp. 522-529, 2018, DOI:10.18535/ijsrm/v6i7.em03.

[79] Y. R. Edward and L. M. Kaban, "The effect of transformational leadership and competence on employee performance with job satisfaction as intervening variable," Acad. J. Econ. Stud., vol. 6, no. 2, pp. 62-72., 2020.

[80] P. Ricardianto, L. Agushinta, R. F. J. Sakti, E. P. Perwitasari, and Y. Pratikno, "Employee engagement: As a mediator of the ship Crew's work effectiveness," PalArch's J. Archaelogy Egypt, vol. 17, no. 6, pp. 7016-7042, 2020.

[81] E. Soane, C. Butler, and E. Stanton, "Followers' personality, transformational leadership and performance.," Sport. Bus. Manag. An Int. Journal., vol. 5, no. 1, pp. 65-78, 2015, DOI:10.1108/SBM-09-2011-0074.

[82] G. Sheppard, "Work-Life Balance Programs to Improve Employee Performance.," Walden University, 2016.

[83] M. Y. Ghadi, M. Fernando, and P. Caputi, "Transformational leadership and work engagement: The mediating effect of meaning in work.," Leadersh. Organ. Dev. J., vol. 34, no. 6, pp. 532-550, 2013, DOI:10.1108/LODJ-10-2011-0110.

[84] W. H. Macey, B. Schneider, K. M. Barbera, and S. A. Young, Employee Engagement, Tools for Analysis, Practice and Competitive Advantage. Chichester, West Sussex, UK: Wiley-Blackwell, 2009.

[85] O. Nazir and J. U. Islam, "Enhancing organizational commitment and employee performance through employee engagement: An empirical check," South Asian J. Bus. Stud., vol. 6, no. 1, pp. 98-114, 2017, DOI:10.1108/SAJBS-04-2016-0036.

[86] M. Hamid, T. Javed, F. Sarfraz, and M. Iltaf, "Impact of employee engagement on organizational effectiveness.," Glob. J. Econ. Bus. Adm., vol. 3, 2018.

[87] F. Cesário and M. J. Chambel, "Linking Organizational Commitment and Work Engagement to Employee Performance," J. Corp. Transform., vol. 24, no. 2, pp. 152-158, 2017, DOI: Linking Organizational Commitment and Work Engagement to Employee Performance.

[88] P. Gupta, S. Steen, and A. Rasheed, "Big Data Analytics As a Tool to Monitor Hydrodynamic Performance of a Ship," 2019, DOI:10.1115/OMAE2019-95815.

[89] P. Pounder, "Responsible leadership and COVID-19: Small island making big waves in cruise tourism," Int. J. Public Leadersh., vol. 17, no. 1, pp. 118-131, 2021, DOI:10.1108/IJPL-08-2020-0085.

[90] A. Radic, J. M. Arjona-Fuentes, A. Ariza-Montes, H. Han, and R. Law, "Job demands–job resources (JD-R) model, work engagement, and well-being of cruise ship employees," Int. J. Hosp. Manag., vol. 88, p. 102518., 2020, DOI:10.1016/j.ijhm.2020.102518.

[91] Y. Bhattacharya, "Employee engagement as a predictor of seafarer retention: A study among Indian officers," Asian J. Shipp. Logist., vol. 31, no. 2, pp. 295-318., 2015, DOI:10.1016/j.ajsl.2015.06.007.